Science-Christianity and Church Activities in the Samoan Islands

Early 21st Century: An Update

FUIMAONO FINI AITAOTO

LifeRich Publishing is a registered trademark of The Reader's Digest Association, Inc.

LifeRich Publishing books may be ordered through booksellers or by contacting:

LifeRich Publishing
1663 Liberty Drive
Bloomington, IN 47403
www.liferichpublishing.com
844-686-9607

ISBN: 978-1-4897-5029-7 (sc)
ISBN: 978-1-4897-5022-8 (e)

Library of Congress Control Number: 2024903630

Print information available on the last page.

LifeRich Publishing rev. date: 03/01/2024

Contents

Acknowledgement

The following have provided various support and assistance for this book: F. Tamasailau Finiana J. Fini; Rennet and Daiana Peseti; Treasure and Jedidiah Levaomana, Alieta Aitaoto, Franny and Toa Halbrooks, Dr. Nia Aitaoto, Joshua Finau, Christopher, James and Penehuro S, Fauloloa Faagata, Seutatia A. Faumuina. Thanks to my IT guys Tepatasi Lealofi and Gideon Finau but especially my incredible wife, the late Daiana Sesepasara Aitaoto.

I would also like to acknowledge my classmates, the Samoa College graduating class of 1973 for their encouragement.

Preface

This book came about because of the need to continue documenting events and progress of the churches in the Samoan Islands as there is no other available literature related to the Early 21. Century. The author's previous book: *Progress and Developments of the Churches in the Samoan Islands: Early 21*[st] *Century: LifeRich Publishing.2021* provided related church updates and it is crucial to continue this almanac-style documentation. A more comprehensive description of the various churches and their related activities were provided in that book.

This book provides a global view regarding Science and Christianity, directing readers to numerous resources like books, articles, papers and most recent aligned discoveries. By science I mean principally theories on the origin of the universe and new archeological discoveries related to Christianity. Plentiful Sources and Footnotes are also presented to not only validate the authors perspectives but mainly to provide resources to Bible College and Seminary students for their researches. (Author's first book has been used by the APTS Seminary in the Philippines in one of its Doctorate courses - Rev. A. Fuimaono AOG).

As of the date of this publication, there were no known similar local literature available.

Related work by Not-for-Profit Organizations, discussions on atheists, televangelists, local politics, climate change, translations, paranormal and culture, in relation to Christianity, are also discussed. Samoan-style Commentaries, with some local contexts are included. The emerging Artificial Intelligence (AI) impacts on many aspects of life is discussed including its influence on local churches. A short sermon regarding creation has been prepared for the book launch, using AI, to illustrate this phenomenon.

Current information in this book regarding science as it relates to Christianity will hopefully encourage Samoan students to study in the field of science, while keeping the Faith, and Christian students is one of the groups that can fulfill this need as these islands are experiencing a serious shortage of Science Teachers, Medical Professionals, Local Engineers and Pastors with some basic science knowledge.

Introduction

The Samoan Islands, are located 14 degrees south of the Equator near the International Dateline. It consists of a chain of thirteen islands, nine of which are inhabited plus two remote atolls. These islands include the islands of independent Samoa which include Savai'i, Upolu, Manono and Apolima, and also the islands of American Samoa, a United States Territory, which consists of Tutuila, Aunu'u, Rose Atoll (*Muliava*), and the Manu'a Islands (Ta'u, Ofu and Olosega). (The elder generations of the Manu'a Islands have a slightly different accent than those of the elderlies of Tutuila islands). Swains Island is also part of the US Territory of American Samoa and its native people have their own language with most preferring that their home island be referred to as Olohega (related to the Olosega island of the Manu'a group).

Because these are small isolated islands in the Pacific, the following list of people with Samoan heritage may help better describe our people. Seiuli "The Rock" – a professional wrestler and actor; Creg Louiganis - Olympic diver gold medalist in the 1984 and 1988 Olympics; Joseph Parker – former WBO Heavyweight boxing champion; Peter Maivia – champion professional wrestler, actor and stunt Coordinator; The Wild Samoans – former Tag Team wrestling champions; and scores of US NFL football players. Judge Tuiloma Neroni Slade is a presiding Judge of Pre-Trial Chamber at the International Criminal Court in the Hague.

New research by the Stanford University published in *Nature* in 2021 appears to confirm that Samoa is the cradle of Polynesia. It is believed Samoa was settled by around 800 BC. This underscores the importance of the Samoan Islands to Polynesia in addition to its warm welcome for John Williams and the first missionaries of the London Missionary Society (LMS) who brought Christianity to our shores on the vessel Messenger of Peace -also known as the Olive branch. A John Williams auxiliary ship sunk at the coast of Avao Savaii in December 1948. Some records noted that Beachcombers and shipwrecked foreigners inadvertently and informally introduced Christianity to the Samoans. Missionary John Williams of LMS considered Samoa as the "most beautiful of the archipelagos of the ocean but also as the most important ones (The Samoa Islands. Dr. Augustin Kramer, Translated by Dr. Thedore Verhaaren Vol. II. Pg.26. Polynesian Press 1995).

As I document the progress of the churches in the Samoan Islands, it's comforting to write this book knowing that Samoans have utilized the foreign concept of Christianity to bolster its community's welfare by fusing it with its exquisite culture. Note, politics is not a contributor to this bliss. Local and overseas churches have undergone modern and sometimes significant reforms in the past two decades. For example, Pope Francis commented (AP 6/14/2022) that, "traditionalists Catholics, particularly in the US, are gagging the church's modernizing reforms and insisted that there was no turning back"

Information about the churches overseas is included since these are the "mother" institutions of all of the local churches. More basic information regarding the various denominations[1] is included in my first two books while this text is **more of an update**. A more global (mainly concerning the US since American Samoa is a US territory) and expanded view about Christianity is relevant for this period as commerce and technology has shrunk distances between countries.

---

[1]

Additionally, members of the clergy should expand their views, interpretations and perspectives towards a more global approach in preaching the gospel, to match the changing contemporary problems and new related archeological and scientific discoveries.

Several essential charitable programs implemented by Not-for-Profit Organizations or Non-Government Organizations (NGOs) are documented here due to their staunch relation to the social work of the churches. Social issues where churches have contributed to are also included.

I had initially thought that a second volume of my perceived series would be published by someone in around 2030, but the significant utilization of my first two books by the religious community including Seminaries and Bible College students, and also the emergence of new church issues prompted the publication of this almanac-style book sooner than professed. The advice from Rev. Filipo Tuigamala and Rev. Lalagofaatasi T. Sanonu to continue with this third book, also motivated me to do my final text on this subject. Because of the modest period covered in this book, readers can refer to my second book (Progress and Developments of the Churches in the Samoan Islands: Early 21st Century) for additional information regarding various churches. My first book (*Tala Fa'asolopito o le Ekalesia Fa'apotopotoga a le Atua i Amerika Samoa (Assemblies of God, AOG)*, was written to document the history of my current church, Assembly of God (AOG), and was written in the Samoan language so the elder generation who helped establish this denomination in American Samoa are able to read and appreciate their critical contributions to spreading of the Gospel of Jesus Christ around these islands.

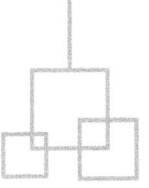

Methodist Church

(Lotu Metotisi)

I n 1828, Methodist missionary Peter Turner arrived and found out about the Samoans haphazard knowledge of Christianity but didn't establish a church. In 1839, Methodism was ordered to be abandoned. In 1857, Australian-based Methodist Church representatives returned to Samoa. Wesleyan missionaries arrived in American Samoa in 1901. Methodist Dr. George Brown was appointed in 1863 to work in Samoa and he worked until 1917. Methodist Church representatives arrived in Samoa from Tonga. The Autonomy of the Methodist Conference was held in 1964. This church currently has its main campus in Faleula, Samoa.

Methodist Church

Caption: Methodist Church in Apia Harbor

The Methodist Church held a ground breaking ceremony during this period for its new gymnasium at the Susana Uesele compound in Ottoville, Tafuna, American Samoa. Church pastors and their wives from throughout the territory and the various church committees joined the President of the American Samoa synod in the celebration.

A new Church building of the Methodist church at Vaimoso was dedicated in June 2022 and was well-attended by several members of the church. The estimated cost of the project was SAT$1, 519,208.07. Funds from donations and a loan from the Methodist Church were used to fund the project as parishioners struggled through the COVID-19 pandemic restrictions. The contractor was paid SAT$330,000 and the new building has air conditioning with sliding doors and windows, and could seat about 400 people.

The Theological College at Piula (PTC), in 2022, announced that twelve successful candidates have passed the annual entrance exam. The candidates included four from Apia Sisifo, two from Upolu Sasa'e, three from Hawaii, one from Upolu Sisifo and two from Salafai Sasa'e in Savaii. According to the College, passing this exam doesn't mean automatic entry to the College. Candidates need to go through other assessments of character and interviews before entering the Foundation Programme which is a one-year programme designed to connect prior learning and the commencement of their theological and ministerial training. The scope of this program covers Ministerial Formation standards and the Introduction to the Academic Programme. Students will then be able to enter the Diploma in water Level 6 Programme. PTC trains people of faith to become agents of change in the lives of God's people, and continue to interpret the Faith through a fruitful collaboration of Academic Competence, Methodist Spirituality and Samoan Cultural Values, according to the PTC Mission Statement.

In July 2022, a former church minister of Safotu village, applied to be discharged without conviction. He was originally charged with manslaughter, actual bodily harm and causing grievous bodily harm to a young man of his congregation who later died. He had entered a not guilty plea.

The church was recognized by the "Beautiful Samoa" campaign in 2022 for its contribution to the campaign's aim to stave off illnesses and diseases and to protect the country's natural environment. Other entities including churches, also received donations from this campaign.

In 2022, the church reappointed Rev. Faulalo Leti and the Ministerial Board to lead the church through 2023-2024. The Ministerial Board include the Treasurer and the Church Secretary. There were several delegations who were not able to fly in from overseas countries as there were still some restrictions on travel due to the COVID-19 pandemic.

During the 2022 Annual Conference, eighty-seven retired mothers at age 70 were honored during a ceremony at Faleula. The ceremony concluded with a family session and the exchanges of gifts.

In 2022, the Methodist Church is American Samoa celebrated new facilities built inside the Tafuna Susana Uesele church compound. The facilities include a hall, gym and a spacious parking lot. The project, which started in January 2021, also included renovations to residences of church ministers and a wall on the perimeter of the compound. Church leaders of other denominations from Samoa and several local dignitaries participated. The estimated cost of the whole project was $3.5 million.

Twelve new ministers were ordained during a Sunday special service ceremony held at Mulivai-o-Aele Faleula temple, at the Church's Headquarters during this period.

In 2023, The President and Board of the Methodist Church Samoa were reappointed in an election during the Annual Conference at the Faleula compound. Rev. Faulalo Ieti Paaga remained as President, Church Secretary Eteuati Epa remained as well and Rev. Ellice Soliola as Church Treasurer. The Conference was also made aware of new churches being established and the continuation of missionary work overseas. Additionally, 13 students graduated from Piula Bible College.

The church at Lotofaga, Safata dedicated a new project on November 2023 that was funded by UNDP, GEF, SGP, Global Environment Facility Small Grant in collaboration with the local government. New trash bins and greenhouses were provided as part of this project that was realized by the church's youths.

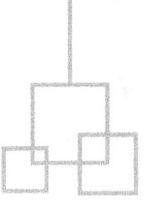

Seventh Day Adventist Church (SDA).

(Ekalesia Aso Fitu o le Toe Afio Mai)

The Seventh-day Adventist church arrived in Upolu in 1899 but earlier representatives arrived around 1890. In 1895, SDA Samoa-Tokelau Mission was established, organized in 1921 and re-organized in 2005.

SDA believe in the following concepts according to Pr. Mark Finley (HopeLives365.1999) a well-known author and an SDA Pastor who had preached the Gospel in more than 50 countries for many years"

1. The Bible is the authoritative Word of God.
2. They believe in the Trinity concept
3. Do not believe in salvation by work and that people are saved by grace
4. Believe in Keeping the Sabbath
5. Its members are a group of people with hope…that Jesus is coming soon.
6. Do not believe that you can eat your way to heaven; human bodies are temples of God; do not condemn people that eat meat, advocate non-smoking, and prefer healthy diets.

In 1907, the First Adventist Church was organized and tracts were distributed. The first church school was opened in 1930. SDA in American Samoa traces its beginning to the arrival in 1944 of the missionaries sent by SDA in Samoa to start up the church. 2022 marks marking 78 years since the arrival of the SDA church in American Samoa. It was during 1891 when the Pitcairn 2 (a schooner built for the Seventh-day Adventist Church for use in missionary work in the South Pacific) arrived in American Samoa for a week before it sailed back off into the South Pacific. After 51 years in 1944, a first missionary from Samoa arrived in American Samoa to start the ministry. He was Pastor Tini Inu Lam Yuen, his wife Fuea and their three young children. Two years later in 1946, the first SDA church chapel was built in Satala. The church slowly grew and after ten years in 1956, nine new churches were built in nine different locations in American Samoa including Alao, Alofau, Masefau, Satala, Nuuuli, Iliili, Vaitogi, Malaeloa and Leone. The first SDA elementary school was established in Satala in 1951 before it was shifted to it new location at Iakina in 1975. In 1957, Pastor Lam Yuen and his family were called back to the work in Samoa. In 2015, leaders of the SDA Trans-Pacific Union in Fiji, along with leaders of the South Pacific Division in Australia agreed to transfer the SDA church in American Samoa to the supervision of the Trans Pacific Union Mission (TPUM) to prepare the church to become a mission. The church celebrated 76 years in American Samoa during this period. The SDA has 21 churches including one church in Ofu, Manu'a. There were about twelve pastors serving the ministry and 22 teachers teaching at the Iakina Academy during this period.

The Seventh Day Adventist church in Samoa hosted their annual "Pathfinders Fair" during this period with the SDA school field at Lalovaea packed with congregation members of Upolu SDA churches. The event was set for all the SDA churches in Upolu but unfortunately not everyone could attend due to unforeseen circumstances. Only 25 churches participated from a total of 60 anticipated to participate. The program was for kids ages 10 – 14 and included activities like building a shelter and camping.

An Adventist church in Oregon, The White City Church

became an official group in 1991; became an Adventist Company 13 years later and in 2009 it was organized and accepted into the sisterhood of churches. In 2016 it changed its name from White City Church to Central Point Samoan Church. In September 2021 the church building was officially dedicated and the mortgage papers were symbolically burned. The church has just over 100 members.

In 2021, a special service was held at the headquarters of the Seventh-day Adventist Church American Samoa Mission at Iakina to officially welcome the new church president, Pastor Kenneth Lelei Fuliese Maisa and his family. It was about three months since Pastor Maisa was appointed as the president of the newly established Mission. The delay in arrival of the new President was due to COVID-19 pandemic restrictions. The special service was led by Pastor Uili Solofa, who was the SDA District Director for five years before the church reached the latest milestone in its history when it officially became a Mission at the beginning of the year. Pastor Solofa said that the new president was appointed by God to come and lead the church. After the ceremony there were traditional presentations. Before being appointed to lead the church in American Samoa, Pastor Maisa was the president for the SDA Samoa Mission for 5 years. In accepting his calling, Pastor Maisa said they would continue to listen and rely on God to give them the strength and courage to face the journey ahead of them. He acknowledged the leadership and the tremendous work by Pastor Solofa and the leadership teams in the past.

During this period, the Seventh-day Adventist Church (SDA) in American Samoa became a Mission. The ceremony was held at the Iakina Church in Iliili and was attended by several government and church leaders. The 92-year-old Siniva Samatau of Vaitogi who was baptized in 1956 was present. President Paster Uili Solofa delivered the divine message while retired church worker Falesoa Puni delivered the history of the church.

The church held a Youth Congress event in December where many young people from its local churches and also from the

Tokelau Mission (STM) participated. The theme of the event was: "Your True Identity". Guest speakers included Trans Pacific Union Mission (TPUM) youth director, Fulton Adventist University College Principal Dr. Ronald Stone, and TPUM President Pastor Maveni Kaufononga. The events included workshops and lectures by the Ministry of Health (MOH) and Samoa Family Health Association. The topics included emotional and physical health, relationship advice and finding a stable future and relationship. An ordination service was also held for two pastors, Atileo Faalelei and Peniamina Ufi.

In May 2023, SDA in collaboration with the Samoa government Ministry of Agriculture and Fisheries project exhibited various agricultural products in its headquarters in Lalovaea. It also commemorated the Church's arrival in these islands 132 years ago.

In June 2023, the Adventist Disaster Relief1 Agency (ADRA) Samoa, received $532,608 from the US Agency for International Development (USAID) to support vulnerable families to reduce vulnerabilities to climate change and promote resilience, economic growth and the safety and well-being of women and children in Samoa. ADRA Samoa will work with women-headed households, low to no-income families, those with disabilities and also abandoned families to build knowledge on issues like financial literacy, agriculture, violence against women and children, disasters and risk management. ADRA is the official humanitarian agency of the Seventh-day Adventist Church.

Professor Glenn. Townend, the SDA Church Pacific Division President stopped over Samoa in 2023 and dropped by to visit the Prime Minister and the Head of State. The visit was to "encourage all the ministries of the church". He also mentioned the great work that their ADRA (Adventist Development and Relief Agency) was doing regarding disasters, agricultural programs and lifestyle diseases. The President was excited about the improvement in structures and the conferral of the first female Prime Minister since his last visit a decade ago.

The SDA Church building of Satomai was saved from destruction by a fire due to the quick action of the Deacon and the village youths in October 2023.

About 200 members of this church from around the world were in Apia for their Annual Health Summit during this period. This Summit was initiated based on the statistics that the number of Samoans to die from Non-Communicable Diseases (NCD) has skyrocketed to 81%. The theme of the Summit was "Step Up to Wholeness" and it encourages people to promote healthy living in order to reduce NCD

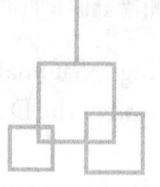

Congregational Christian Church of Samoa (CCCS),

(Ekalesia Faapotopotoga Kerisiano i Samoa, EFKS)

The London Missionary Society (LMS) missionary John Williams, accompanied by some Tahitians, arrived in Savaii in 1830 bringing Christianity to these islands. John Williams and Misi Harisi remains were later returned to Samoa via a British warship. Between 1832 and 1836, LMS missionaries visited American Samoa. An LMS church was later established on Tutuila in 1838. The Malua Printing Press was established in 1839 and 12 LMS Missionaries left to work in Melanesia during the same year. In 1844, the LMS Malua Theological College (MTC) was established in Samoa. LMD also established the Leulumoega High School in 1890 and the Papauta Girls School was founded in 1892. The famous writer Robert Louis Stevenson gave a speech at MTC in 1889. The Samoan version of the New Testament was printed in 1848 and the Old Testament Samoan version was printed in 1855. In the same year, the LMS Fagalele Boys School was established in Fagalele, American Samoa. John C. Williams, son of missionary John Williams became the British Consul in Apia in 1858. In 1871, the second group of LMS missionaries arrived in the south coast of Papua New Guinea. An old article titled *"O le Aso Malolo e manatua ai oe"* about the work of

Samoan Missionaries in Papua New Guinea was located by the "*O le Sulu Samoa Research Project*", and was written by my father (Rev. Aitaoto Seiuli Fuimaono) while serving as a missionary in Papua in 1948, and translated by W. Ieremia Allan.

Malua Church

Caption: CCCS Malua Faleupeli Church

A new church was built in 1892 in Apia (a'ai o Niue) on land donated by Chief Seumanutafa Pogai. In 1900, the LMS Atauloma Girls School on Tutuila was dedicated (the school was initially erected in Amanave). A school was opened in Tau, Manua in 1908 with only one teacher, from LMS. In 1962, the LMS became the Congregational Christian Church of Samoa – Ekalesia Fa'apotopotoga Kerisiano i Samoa (EFKS). Theodore Mila Sapolu became the first Samoan to be selected as the Principal of the Malua Theological College. Additional information regarding this denomination can be found from texts by Rev. Elia Taase and Rev. Oka Fauolo.

Samoa Government's statistics from its 2021 Census confirmed that CCCS is still the largest denomination in the country followed by the Catholic, LDS and Methodist. The observation that some congregational denominations have lost some of its members to other denomination is true but the peaceful Sunday atmosphere in the villages continue to reflect the continuous attendance of local people in churches. Some SDA members occasionally use public places for parties and games on Sundays.

Church ministers of the Congregational Christian Church of Samoa held a four-day spiritual retreat (Mafutaga Faafouina) at Malua in 2021. The retreat was to enable church ministers to meet and discuss topics prepared by the Principal and teachers of Malua Theological College. The gathering is an annual meeting for the church ministers to encourage them in their journey as they strive to provide the service of preaching the Gospel in different parts of the world. Overseas members weren't able to attend this retreat as well as the previous year's retreat due to COVID-19 restrictions. Overseas members however were able to use the Zoom platform to join in this year.

During this period, the service of Reverend Victor Pouesi of the Congregational Christian Church of Samoa (CCCS) in Mangere East to the local Samoan community earned the rare recognition of being a Queen's Service Medal winner in New Zealand. A special medal was awarded by the New Zealand Government to recognize his outstanding volunteer service to the community since 2002. Reverend Victor and his wife Salome Pouesi were called to serve at Puaseisei Mangere East church in 2002. Pastor Pouesi also played a critical role during the COVID-19 outbreak. in South Auckland. Pouesi initiated the Taeaofou-i-Puaseisei Playgroup, a bilingual Samoan center established in 2006. The Center provides opportunities for children to embrace their Samoan culture and heritage while creating local employment opportunities.

In December 2021, the Congregational Christian Church of Samoa (CCCS-EFKS) at Afega dedicated its multipurpose hall with the new building now called the ʻAfega o Alofa[2]ʼ hall. The multipurpose hall was built by the Zheng Construction Company and took eight months to complete. The total cost of the project was approximately $956,700 and took eight months or 31 weeks for the construction works to be completed. The multipurpose hall is 40 meters in length, 15 meters in width but 8 meters high. The church minister in charge of CCCS Afega is Reverend Luapene Nepo and his wife Faaolataga Nepo.

CCCS Safotu dedicated its $334,579 new hall in May 2021. CCCS Chairman Rev. Elder Iosefa Uilelea witnessed the event while Tekimatang Uilelea was given the honor to cut the ribbon.

In November 2021 former students of MTC donated $50,000.00 worth of equipment to the school's library. The equipment would help improve the learning environment for the "heart" of the biggest denomination in the country. The pastors noted that this type of project will nurture the lives of future ministers.

During this period, the Malua Theological College, owned by Samoa's biggest church denomination, Congregational Christian Church Samoa (CCCS.) announced that it was ready to offer a Master's in Theology degree program. This was confirmed by the Principal of Malua Theological College, Reverend Dr. Vaitusi Nofoaiga in an interview with the Samoa Observer. He said the initiative to introduce a Master's in Theology degree program at the theological college was signed off by the Elders of the CCCS and members of the 2016 *Fonotele* (Annual Conference). The delay in the implementation of the program was due to the lack of resources available in the school. Nevertheless, Reverend Dr. Vaitusi is confident that the staff of the college are now ready to start the program. Various resources are still needed to implement this program but the staff were ready to start the courses in 2022. "Anyone who came through MTC and is now a church minister is eligible to apply for our master's program next year. Around 2021, there were more than a hundred students enrolled at MTC. under the supervision of thirteen (13) teachers, with two returning teachers who just graduated with their doctorates from Australia.

Another related project was the launching of the MTC "Samoa

Journal of Theology which will include multi-discipline academic writings by the teachers of Malua Theological College. The Samoa Journal of Theology according to Rev. Dr. Vaitusi has been approved and recognized internationally, with renowned overseas Professors of Theological studies and historians who have offered their help with the publication. "We have professors from Australia (Professors Mark Brett), America, just to name a few and also Professor Leasiolagi Dr. Malama Meleisea. MTC still has the Malua Journal which is written in Samoan. MTC also has an online program called EBCSO eBooks which the students can use to get any books they need for their research and assignments. I hope they would update the church's website regularly utilizing this new knowledge.

The Puapu'a CCCS dedicated the last service of the year to its main choir in recognition of their hard work throughout 2020. This is an Annual Event and has been practiced by the church for 40 years.

Reverend Melepone Isara, currently a Lecturer at Malua Theological College, was appointed to the position of Treasurer during the Congregational Christian Church of Samoa's Annual conference. The appointment of Reverend Melepone was endorsed on the final day of the annual conference, one day after his ordination as a church minister (*Faifeau Samoa*) on Saturday.

Reverend Melepone is taking over the role after the former CCCS Treasurer, Reverend Rimoni Wright, who resigned from the post to serve as a resident minister at Matautu Falelatai.

Melepone was happy since he was not only ordained on Saturday, but the church has also gained their trust and voted him as the new treasurer for the church. Reverend Melepone has been a lecturer at Malua Theological College since 2019, after graduating from the Pacific Theological College (Fiji) in 2018.

Circa January 2022, the CCCS at Patamea, Savaii church was destroyed by fire. According to a Global News report, the church building was completed in 2004 and the congregation has just paid off their associated loan. The congregation immediately started to fundraise for a new church building and the FAST party gave its support to the fundraising because its Deputy Prime Minister is a member of this congregation. The congregation has 41 active member families (*matafale*)

In early 2022, EFKS of Apia welcomed its new pastor, Rev. Dr. Latu Latai. Traditional presentations followed at the pastor's official residence, a building that I used to attend Sunday School at in the past as a young kid from Apia. Traditional presentations included a $50,000 gift to the pastor's family. The previous retired pastor was Rev. Utufua Naseri – married to my classmate Taiaopo.

In May 2022, the CCCS Church of Tafitoala welcomed their new pastor, Reverend Tanuri Pasaleli in a ceremony led by Rev. Elder Kereta Fuafiva. It has been nearly a year since the church was without a pastor. The pastor served at the Aufaga and Alafua CCCS before this calling. Faletua Sanaima is the daughter of Rev. Elder Tavita Anisone.

During this period, the CCCS re-scheduled its *Fonotele* to September 19-30, 2022 due to COVID-19 restrictions. These meetings are normally held in May. The new dates were declared after the Samoa government planned to reopen its borders in August 1, 2022. The *Fonotele* is the main general meeting where the budget and resolutions are passed.

A new house of worship, under the leadership of Pastor Koro was opened during this period for the CCCS at Saleaula. Many church and government leaders attended the event including the Speaker of the House, Acting Prime Minister and CCCS Fonotele Chairman. Construction was done by the Evaeva Construction under stressful conditions due to COVID-19 restrictions. The whole project cost about $1,120,896. This village is part of the Gagaemauga No. 1 electoral constituency. The old church building was demolished in August 2021.

In July 2022, the CCCS of Falelatai, held a special service to dedicate a new plaque commemorating the establishment of the first printing press in the Samoan Islands. The first printing press was established in 1839 at a piece of land called Matanofo, Falelatai, that was donated to the then LMS church by Chief Tuimalealiifano. My father (Pastor Aitaoto Seiuli Fuimaoano) worked at this Printing Press when it was located at Malua.

In August 2022, the CCCS of Apia celebrated 130 years of its church building, *Diamond of the Ocean*, which was built on August 31, 1892. This building is significant because it is the final resting place of the renowned LMS missionary, Reverend John Williams, whose remains are buried underneath the entrance into the church. According to Rev. Clarke and Rev. Goward of Apia LMS Mission, this church was once known by a few names including the *Apia Native Church, Centenary Church, John Williams Memorial Church*. The ordaining of pastors travelling to Papua as LMS missionaries (including my parents) took place in this church[3]. John Williams brought Christianity to the Samoan islands in 1830 and was later killed in Eromaga (now Vanuatu). A John Williams Memorial is across from the church on the Apia waterfront.; where I used to sit and play with my young friends, many years ago. It is a great honor to include this information in my book, as a former member of this church and its Sunday School. Some photos and information regarding the history of the church was provided by the *Te Papa Museum* in New Zealand and can be found on the EFKS Facebook Page. Following is a list of pastors who served in this church:

- Reverend Nemaia & Noema (29 Nov. 1885-1889)
- Reverend Solomona & Senetima Esene (1889-1918)
- Reverend Pouesi & Siuila (1918-1942)
- Reverend Tapeni Ioelu – 1944
- Reverend Faulalo and Liai Sagapolutele (1958-1996)
- Reverend Galuefa & Sooletala Aseta (1983-1996)
- Reverend Utufua & Taiaopo Naseri (1998-2019)
- Reverend Latu & Lotu Latai (current).

3

This list includes Reverend Pouesi who left the church after an argument with some church leaders, and formed his own church which was then known as "Lotu Pouesi". My father (Reverend Aitaoto Seiuli Fuimaono) was one of the Apia LMS parishioners who left with Pouesi to form the new church. This church spread to American Samoa and is now known as the Congregational Christian Church of Jesus in Samoa (CCCJS) – *Ekalesia Faapotopotoga a Iesu I Samoa*. Some of the American Samoa CCCJS Pastors include Rev. Ueli, Rev. Levi, Rev. S. Maualaivao, Rev. Lauoi Mageo; Rev. Faamao Asalele, Rev. Emau Amosa and Rev. Patolo Mageo.

During the 130 years anniversary celebration event, the story about a fire that broke out in the church in 1917 was cited by the Church Minister. Chief Seumanutafa was told about the fire and he told the people of Apia to save the church even if its cost them their lives. This was a touching story and reflects the peoples love for their house of worship. A *taulaga* collected more than $181,000 tala and a special song composed by Asiata Melvin Solomona specifically for this occasion, was sung by the choir. A special plaque to commemorate the 130th. Anniversary was also launched.

In September 2022, CCCS reviewed its constitution and compiled amendments to be submitted to the Annual General Assembly- *Fono Tele*) the following week (this was around the same time American Samoa reviewed its constitution and solicit public suggestions for needed amendments)[4]. The Review Committee started its work on September 12, 2022 at the Malua Meeting House. The usual Annual Conference was disrupted by the COVID-19 restrictions but Samoa finally opened its borders for overseas travel in August 2022. One of the topics that was to be discussed in the following conference would be the election of the Church's Secretary General. According to recent statistics, the church has more than 70,000 members. The church has recently celebrated 192 years of

---

4

the arrival of Christianity at Sapapalii, an event that attracted many Samoans residing overseas.

In late 2022, CCCS's Malua Theological College announced that 28 entrants passed the annual entrance exam for 2021. The number had increased to 28 from previous year's 23. The new entrants included six candidates from New Zealand, twelve from Samoa, two from Hawaii and eight from Australia. These candidates are required to complete a *Ta'utinoga* – a stringent interview process conducted by the High Council of the church, the *Komiti o Toeaina Faatonu,* to "test character, attitudes and former behaviors". Candidates are also required to provide a medical certificate to ensure they have a clean bill of health.

On the 9th July 2012, a one-day workshop was held by the CCCS Mision Office to address alcohol and drug abuse. Participants included eight members from each respective CCCS College namely Leuilumoega Fou Collge (LFC) Tuasivi (TSC) Maluafou (MFC), Nuuausala (NSC), Paputa (PGC) and the Congregational Senior College (CSC).Other participants were from different denominations. Presenters included employees of the Health Department, the Ministry of Police, CCCS Counselling Department, a young student Motivational Speaker and a lecturer from the Malua Theological College.

During this period, Rev. Elder Iosefa Atapana U. passed away after more than 30 years serving in various notable capacities within the church. He entered Malua Theological College in 1985 and served as pastor for the Lotofaga EFKS in 1989. Samoa's Prime Minister is a member of this church.

A community project conducted by the church called the EFKS Vini Fou Conserving our Natural Spring and Waterways Project, restored and rehabilitate the village's freshwater pool called *"Punaoleola",* in early 2023. The project cost $55,000.00 Tala and was funded by the United Nations Development Environment Facility Small Grants Programme (GEF-SGP). This is another good example of how churches contribute to environmental issues.

A new Church building was dedicated in February 2023 for the EFKS at Patamea, Savaii, under the leadership of Rev. Tanielu Tupuse. A previous church building was destroyed by fire in 2012 and the parishioners immediately started fundraising to build this new church building. The original church building was completed in 2004.

In early 2023, it was announced that a Samoan lady, Wanda Ieremia-Allan, was heading to the Cambridge University to research the historical legacy of the legendary publication *"O le Sulu Samoa"*. This was the first newspaper published in the Samoan language and was the work of several people including, grassroot journalists, tranlators, diarists, travel writers, book binders and printers. The publication mentions the Samoan Calendar, traditional metric systems, natural weather patterns, church events and much more. This Ph.D candidate now has the largest

Digital archive of the famous publication.

Before the start of the 2023 *Fonotele,* the *Mafutaga Aoao a Tina*'s charitable arm, *Tautua Puapuaga Tagata* (TPT) started its meetings. In attendance were more than a thousand wives of pastors, elderly deacons and pensioner mothers.This organization donates thousands of Tala to various entities at the end of their Annual Meetings. Last year, TPT pledged T$70,000 to the Government of Samoa with $10K going to the Samoan National Hospital, $10K for hospitals in Savaii $30K for the Motootua National Kidney Foundation and $20K for the Savaii National Kidney Foundation.

Rev. Auatama Esera of the EFKS of Sapapalii, Savaii, was elected in May 2023 to be the new Vice Chaiman of the EFKS. He took over this position five months after the passing of his predecesor, Rev. Iosefa Atapana Uilelea.

EFKS's General Purpose Committee in May 2023, agreed to ask the Samoa Government to refrain from discussing the same-sex marriage and transgender issues. This issue came up during a discussion in the Samoa Parliament during the first-ever Commonwealth Parliment Association (CPA) branch meeting during this period. HRPP noted that it accepts Samoa's "third gender". According to Prime Minister Fiame, "it is still illegal for people of the same sex to have intimate relationship" in Samoa.

In May 2023, a couple of disticts asked the EFKS Excutives for an update on church funds that a former Assistant Treasurer (Fata Ueligtone Malifa) of New Zealand parishes allegedly used "illegally" during his term of office. The districts needed to know if the funds used by the former Treasurer were recovered. A former general secretary of EFKS informed church members that the accused's brother, the late Rev. Elder Taeipo Malifa (my Samoa College classmate) has started re-paying the funds taken by his brother. The Chairman and Treasurer of EFKS had travelled to New Zealand in 2019 to investigate the matter but it was reported that Fata had moved to Australia.

After serving his church at Manunu for for 31 years, the 70-year old Pastor Roma Enosa bid fairwell to his beloved congregation. He was the fifth pastor for this congregation. The sad occasion included a SAT $200,000 gift for the departing pastor.

A EFKS pastor appeared before the Samoa Supreme Court during this period for criminal mentions. The pastor pleaded not guilty to all five charges filed against him including allegations of fraudlently signing of church funds. The courtroom was packed with members of the church from Savaia some supporting the pastor. It was evident that this event has divided the church according to one parishioners who attended the criminal mentions. The pastor remained on bail with the condition of signing in with police on a weekly basis.

To end this section I would like to offer my personal opinion regarding a couple of practices by this denomination that some people interviewed for this book respectfully dispute.

i.   The reading aloud in church of amounts of money individuals/families donate to the church for various projects and purposes.

A family or individual donating ten dollars would feel belittle or demeaned when another person or family donates one hundered dollars. Most EFKS pastors would not take this issue seriouly since all of them live comfortably in lovely ministers houses. Sitting in LMS churches in the early 1960's, besides friends whose families barely afford the basic necesseties, instilled in me the continous discomfort when this occur in nearly every Sunday service. This observation is not from a text book but from several parishioners'perspectives.

ii. Required amounts of donations are usually made across the border, regardless of a family's household income. This is not only unfair but also un-biblical. Parishioners don't complain out load but swallow their discord in silence. In the past decades, this sometimes results in calling relatives abroad to send over money, leave for another denomination, or stay home and not attend church.

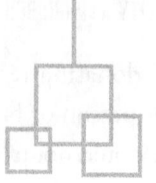

Congregation Christian Church of American Samoa (CCCAS)

(Ekalesia Faʻapotopotoga Kerisiano Amerika Samoa)

T he Congregational Christian Church of Samoa has declared Sunday, June 6 as "Pray for American Samoa Sunday." All member churches are asked to remember and pray for Samoa during their worship/communion service this Sunday. A letter from the General Secretary of the church, Rev Nafatali Falealii said that the church leadership as well as the Committee of Elders has decided to set aside this Sunday as a day of prayer for the people and government of American Samoa. He said, "The situation facing Samoa with its government is known to all. Let us unite in prayer for justice and peace for Samoa during this difficult and trying times." The CCCAS is the territory's largest church with branches in American Samoa, and across Hawaii and the US mainland, New Zealand and Australia.

Seetaga Church

CCCAS Nua Seetaga Church

When the Congregational Christian Church of American Samoa was gearing up for one of its biennial general conferences during this period, COVID-19 struck. General Secretary of the church Rev Nafatali Falealii said due to the Covid-19 pandemic, the conference, that was planned for July 11-23rd will be held via Zoom and there will be no overseas delegates traveling to the territory for the two-week meeting. Local delegates will gather at the "*Ua Taunuu*," church at Kanana-fou for the daily Zoom sessions. The Secretary explained some of the topics that the delegates was to discuss. The work projects being proposed include new housing and roads at Kanana-Fou as well as an upgrade of the gymnasium. There are 350 local delegates and 378 delegates participating via Zoom from Hawaii, the US mainland, New Zealand and Australia. The CCCAS conference is usually a boost for the territory's economy as the parishes buy food and supplies for the delegations from the various member churches, and the overseas delegates love to shop and patronize hotels and restaurants. Churches doesn't only take in money; it occasionally contributes significantly to the community's economy.

CCCAS 2022 Conference in American Samoa elected new officers. These included a new Chairman (Rev Elder Leatulagi Faalevao) Vice Chairman Rev. Faaitete), Treasurer, Secretary (Rev. Nafatali Falealii), Chairman of the Elders Committee (Falelua Lafitaga). Samoa News however, noted that the individual accounting figures provided to the conference were allegedly not consistent with the totals for collections and expenses. This is a very serious matter and must be dealt with as soon as possible so the faithful parishioners wouldn't lose faith with the church. Alleged suspicious handling of funds should never happen to church entities.

On July 22, 2022, retired Reverend Elder Eveni Mamoe Eveni Jr. passed away in San Jose. He was 75. He has served his spiritual duties for 31 years. He was born in Tau, Manua and graduated from Samoana High School and Kanana Fou Seminary. He was a deacon of the First Congregational Christian Church of Samoa in Sunnyvale and was the Elder Minister for the Ituau District of the CCCAS.

This church was invited to participate in the American Samoa Flag Day in April 2022 and was accepted by the Church.

Funding of about $1.2 million was received from the government to perform many major repairs and maintenance work of the 'Maota o Tupulaga – Taeaoafua".

The accreditation of the church's Kanana Fou High School by the Wetsrn Association of Schools and Colleges was completed towards the end of 2021.

Circa late 2021, CCCAS appointed Attorney and Deacon Talaimalo Marcellus Uiagalelei as Legal Advisor to the church. He is a parishioner of the Futiga CCCAS.

Around August 2021, CCCAS paid off a $2 million loan from the UCC Cornerstone. The funds were for repairs to the church building "Ua Taunu'u" after it sustained damages from Hurricane in 2017.

The Annual CCCAS *Fono Tele* XXX111 was held via internet Zoom around July 2021. Financial Reports were presented and reflected good financial standing of the church. The meeting participants also agreed to have the current leadership continue their work and in their current positions until the next *Fono Tele*. At the end of this meeting an additional $161,359.0 was collected through traditional *siva* by several of the churches groups and organizations. Deacon Tuaolo F. also made a substantial donation for the church. Several church committees also met via Zoom in July 2021 including the Women's Executive Committee.

In April 2021, several pastors who have served in their respective churches for several years were ordained and were now ready to serve as pastors for the respective village churches. These included some from the Manu'a Islands.

In June 2023, the Kanana Fou Theological College celebrated its 40th. Anniversary. Several groups from the student body performed various items through the week's event. According to church officials, it was decided that the College wouldn't wait until its 50th. Anniversary but opted to celebrate the 40th. Anniversary so that members of the Class of 1987 could take part. Some members of this class have passed on or have retired. There have been more than 400 students who graduated from the college since the College's founding more than 40 years ago. The Commemoration Celebration replaced the Church's Annual Conference for 2023. The 2023 graduation had 17 students and it was the college's 37th. Commencement Ceremony. At the end of the celebration, the Alumni announced their special gift of a new building for the college.

An ordained lay preacher of the Fagalii church in American Samoa drowned during a fishing trip in 2023. Fortunately, the daughter and her husband who were fishing with the pastor, survived.

Towards the end of 2023, Rev. Elder Dr. Leatulagi Faalevao was elected to lead the Pacific Council of Churches. This was the first time ever a pastor from American Samoa was elected for this important position. Rev Faalevao was the Chair of the Congregational Christian Church of American Samoa when he was appointed to the position of moderator during the 12th. General Assembly held in Noumea. Pastor Faalevao was also the Chairman of the Pacific Theological College in Fiji and had also served on the Central Committee of the World Council of Churches. He has faithfully served his church in Amouli, American Samoa for about 30 years.

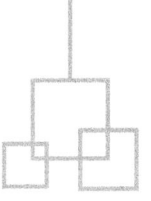

Voice of Christ

V oice of Christ Lighthouse Temple was dedicated in 2021 at the village of Aua. The first Lighthouse Temple was built on *"Mauga o Pioa"*.

In May 2023, nine members of the Voice of Christ Full Gospel completed the Basic Sewing workshop conducted by the American Samoa Community College's Agriculture, Community and Natural Resources Division's Family Consumer Science Program. Participants learned the fundamentals of sewing.

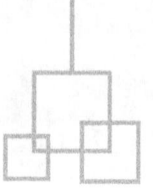

Worship Center Christian Church

I n April 2022, the new "Jesus Christ Worship Center Christian Church Worldwide" three-storey building was dedicated at Sogi, Samoa. The project cost around $20 million Tala and sits on prime land bought via an ANZ $2.2 million loan 20 years ago and assisted by a loan from Andy and Leone Forsgren. The ceremony was attended by the Prime Minister and several dignitaries. The building, build by Ah Liki Construction Ltd., is sound-proof, has seating capacity for 2000 worshippers and the second story is all glass. Work began on the building in 2018.

In April 2023, apostle Viliamu Mafo'e, leader of the Worship Center Christian Church passed away after a prolonged illness. Mafo'e established the Worship Center in 1997 and was instrumental in the construction of the new church Center in *Sogi*. He was one of the few local pastors that preached the Prosperity Gospel and the Y2K event that predicted the end of the world in year 2000. RIP Sir -*Ia manuia lau malaga.*

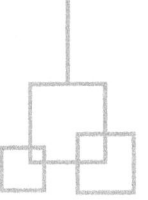

The Zion Christian Mission Centre

This relatively new church appeared during this period and it's part of a mega-church in Korea called the Shincheonji Church of Jesus. Its Regional Director is Mareko Taafua. Samoa media reported (Samoa Observer)stories published by Stuff in New Zealand around November and December 2022, describing testimonies from former members who described how they were "trapped in a cult" as members of this church. Media comments from the public were mixed. A cult is sometimes defined as "a relatively small group of people having religious beliefs or practices regarded by others as strange or as imposing excessive control over members". In the past, many members of these groups end up committing suicide, loose personal properties or be killed.

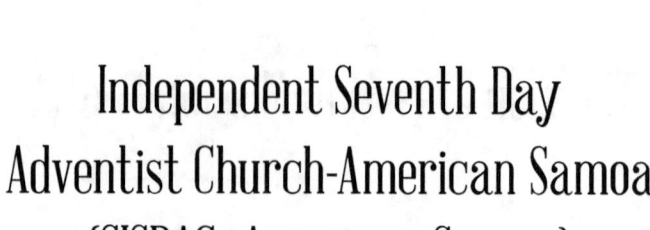

Independent Seventh Day Adventist Church-American Samoa (SISDAC- American Samoa)

(Ekalesia Aso Fitu Tutóatasi)

T his church is reportedly has no association with the Seventh Day Adventist Church in New Zealand and is also not recognized as part of the official Seventh-day Adventist Church.

Circa July 2021, Youth members of the Samoa Independent Seventh Day Adventist Church-American Samoa held a Youth Camp at the church's compound at Ottoville, American Samoa.

A Master Guide ceremony was held in 2022 by the Master Guide Su'e Ala Club of the SISDAC- AS Division in American Samoa.

According to RNZ (Smith, Mackenzie 2-12-21), "the church in Auckland has been struck off the Charities Register after its employees stole millions of dollars in church funds and poured millions into dubious investments". This came about after a two-year probe into the church which operates 10 churches in the Auckland area. An investigation Report by Charities Services detailed mismanagement of millions of dollars of church funds mainly by the Executive Director's relatives and associates.

This Church's Executive Director Pastor Papu had publicly said that he will be martyred in Apia because he has told the truth about the erroneous doctrines of other churches. Then Prime Minister Tuilaepa, the National Council of Churches (NCC) and many churches in the Samoan Islands publicly voiced their disgust towards Pastor Papu's public comments. Maybe this event provides proof of the NCC and the Prime Minister's opposition to Pastor Papu's fictitious personal interpretation of the Bible. I wrote in my previous book that I will visit the leader of this church, Rev. Papu in about five years to check if anyone has hurt him as he alluded to during that highly controversial sermon. I predicted that no one will hurt him due to his controversial sermon and that he will live for at least another 5 years. Missionaries are often killed when they proclaim the truth and are usually not hurt when preaching the opposite.

In March 2023, this church bought the Monalisa Hotel on Upolu. The Hotel had been on the market for four years.

This Church took the Samoa Development Bank to Court in 2023 on a matter regarding a hotel named Monalisa. It was reported that the Church took over the operations of the Hotel in February 2023. The Hotel originally opened for business in 2009.

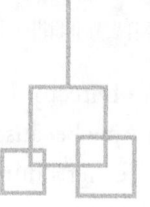

Islam

According to Asma Afsaruddin, Professor of Islamic Studies and former Chairperson, Department of Middle East Languages, Indiana University (Aug. 14. 2021), one day, in Mecca, the Prophet Muhammad dropped a bombshell on his followers: He told them that all people are created equal.

"All humans are descended from Adam and Eve," said Muhammad in his last known public speech: "There is no superiority of an Arab over a non-Arab, or of a non-Arab over an Arab, and no superiority of a white person over a black person or of a black person over a white person, except on the basis of personal piety and righteousness.". In this sermon, known as the Farewell Address, Muhammad outlined the basic ethical and religious ideals of Islam, the religion he began preaching in the early seventh century. Racial equality was one of them. Muhammad's words jolted a society divided by notions of tribal and ethnic superiority. Chapter 49, verse 13 of Quran states: "O humankind! We have made you... into nations and tribes, so that you may get to know one another. The noblest of you in God's sight is the one who is most righteous." Historical inequalities within this community were based mainly on kinship and wealth. Women, declared to be the equal of men by the Quran welcomed Mohammed's message appealing. However, the potential of gender equality in Islam would become compromised by the rise of patriarchal societies. Confusion sets in when considering the way women are treated in addition to being "righteous" by using humans to deliver deadly bombs.

The 2022 discovery[5] of an ancient Christian monastery on an island off the coast of the United Arab Emirates, which possibly dates back years before Islam spread across the Arabian Peninsula was also an interesting new find during this period. Carbon dating revealed it was built between 534 and 656. Islam's prophet Muhammed was born at 570 and died in 632.

There have been recent questions after new discoveries regarding Mecca; the observation that it is mentioned only once in the Quran and the realization that its faithful should be facing Petra when praying (Dan Gibson, HChannel).

The Quran was believed to be written more than 200 years after Mohamed died and was written by people who never saw or listen to Mohamed. This is in contrast to two of the Gospels that were written less than a hundred years after Christ's resurrection. Two of the Gospels were allegedly written by disciples who were actually with Jesus and heard his actual words and teachings. The other two were written according to **first-hand** testimonies of people who were with Jesus.

Muslim Sheikh Khalid said on television that Jesus should have explicitly say that he is the Son of God. Was everything documented during Biblical times or did the writers sometimes fail to mention some events and facts even though the events actually happened and certain words were actually spoken?. My granddaughter (Baby Daiana) said that if the Gospel writers had used PC's or Macs with the latest version of Word and also using Smart phones to record conversations, everything regarding Christianity would have been thoroughly recorded, and maybe the Gospel writers would have a record of Jesus saying He was the Son of God. Reading about Sheikh Khalid's life, he seems to be the only human who never forget anything in his life and was able to document every single

---

[5] AP. Gamberell, Jon. 11-3-22.

event relating to his life and his religion. NOT. He also said that when one goes into a funeral service, one would only hear crying and sighing. Its apparent he never attended a Pentecostal Church funeral service where the clapping, singing and rejoicing can be overwhelming, like in my AOG church in Pago Pago. Khalid you are welcome to join our funeral services anytime. If he was a true Muslim he would entertain the concept of true love – compared to materialism – empathy, compassion and the yearning of the human soul for a deceased person.

A relatively recent comparison between Jesus, Mohamed and Christianity versus Islam done by John Akerberg and Jay Smith suggested the following and I suggest that people interested in this issue should examine the following:

Qur'an: Surah 19:20, 10:94; 21:7' compare this to Isaiah 7:14

Surah 3:46; 3:49, 19:19 - comparing these to Biblical passages since it may suggest that Mohamed couldn't heal anyone; that Jesus didn't sin and that the prophets in the Qur'an sinned. Qur'an can't be interpreted as oppose to the Bible which can be interpreted. It was also pointed out that '*issa*' is mentioned in the Qur'an 93 times while Muhammed is mentioned only 4 times. ":*yasu*" is the Syriac name for Jesus and that the Qur'an has the wrong Jesus, not the Biblical Jesus. The presentation also noted that the word Allah is not an Arabic name and that Islam is a 9th. Century invention.

Some Muslims ask "Can the Christian God die?. Of course. Jesus said "I lay down my life so that I may take it again....I have the authority to lay it down and authority to take it up again: John 10: 18.

According to the above assessment, Christians who have Muslim friends who ask questions regarding the Bible should know that most of the answers to their questions are already in their Qua'ran (Dr. Jay Smith). Atheists are often dissatisfied with Christian's arguments for their beliefs because they say that, Christians' supporting evidences are solely from the Bible scriptures, a text they don't believe in. Well, readers are encouraged to read "Cold-Case Christianity" with J Warner and Jimmy Wallace, where many evidences, not from the Bible, support Bible stories can be found. Additionally with the discovery of the Dead Sea scrolls, doubters can examine related information provided by Dr. James D. Tabor who had studied these scrolls for about 30 years. Interested students can also check out his related course on: http://www.mythvisionpodcast.com/dss.

Christians may also want to study the Qua'ran in order to help their Muslim friends. This is also in the Quran. Surah 10:94; 21:7 "But if you are in doubt as to what We have revealed to you, ask those who read the Book before you". Another example, Muslims believe their Allah is in heaven and not on Earth. Christians believe that their God is in heaven and also on Earth which apparently can be proven by the Qua'ran! Read Surah 20: 10-14. Since no one can take the name of Allah, therefore Allah was in the burning bush with Moses – right here on Earth. Interesting also is that the Muslim's Garden of Eden is in heaven and their Allah is not there. Furthermore, the Qua'ran seem to have borrowed this story from the Bible, and as with other texts, it was borrowed **without context**. Some Muslims have asked if the Christian God can remain sinless. Again, the answer is in the Qua'ran (Surah 19:19). (the angel) said (to Mary): "I am only a messenger from your God, (to announce) to you the gift of a righteous (**sinless**) son". Recent analysis of other old Qua'ran-related texts has revealed that the current Quran has several variants, some wrongly dated events and has many changes[6] like insertions, erased texts, and overwritten passages. Texts proving

---

[6] Jay Smith. Nicene International. 2014

these changes were provided by Jay Smith. To be fair, Jay Smith also offered and acknowledged that some changes were also made to the Bible, a claim that Muslims are adamant that it can't be applied to the Qua'ran. He also mentioned that the Qua'ran was properly compiled in 1924 and was officially adopted in around 1985. After these evidences were presented, Dr. S Ally, a Muslim debating Jay Smith, refused to discuss the origin of the Quar'an but instead went ahead and provided a peculiar mathematical proof of the Qua'ran, then he said that the Qua'ran has the DNA of God. Let me remind Dr. Ally that there is something called "recessive genes" that sometimes doesn't reveal itself but its real. Stay in your lane and figure this out. Additional related arguments can be found in several podcasts by the intelligent Sam Shamoun.

Christians should also be aware that Muslims often ask "Can your Christian God do this or do that etc.", Well of course, HE can, HE is omnipotent?, but there's a catch, HE can't lie.

A Muslim turned Christian said that the two concepts that turned him to be a Christian are the passage in the Bible "love your enemy ......" and the promise by Jesus that he will return for his believers. He said, muslims have an opposite resolve to the love concept mentioned above and that Mohammed is not returning to get his people.

For a more complete analysis of the history of Islam, I encourage anyone to also read "The two Faces of Islam" and watch presentations by Armen A. Saginian in addition to the Jay Smith' analysis above.

After more than ten years of research, I can conclude that Muslims and Christians **don't worship the same God.** Muslims don't believe in Christ and the Trinity and the Quran mentions loving your neighbors and doesn't mention loving your enemies.

In Summary, Readers are however encouraged to continue studying and do research on Religion to get their own opinions and **always keep an open mind.**

In Samoa, about ten years ago, some Muslims were living in the area called Malifa near the town of Apia.

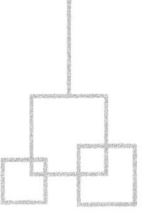

Peace Chapel

Peace Chapel B has claimed the under-15s Boys title for a 3-day cricket competition that saw three private schools battle it out for the top spot. Peace Chapel Primary School's B team took top spot in the Boys division after defeating seven other teams, while Mt. Zion Primary School's B team were the year's U15 champs for the Girls division. Samoa International Cricket Association representative, Ms. Kalala Tanuvasa stated yesterday that during the running of the tournament, they saw a high level of skills displayed by the participating schools. "There were 4 schools initially planned to vie in this tournament. However, Vaiala Beach School was not able to field a team. Even so, we are very pleased with what we've witnessed from the competition, it has improved from previous tournaments. When international cricket was first introduced to schools six years ago, kids were slow to pick up on how to play. But as the years progressed, we have seen a lot of improvement not only in the participation numbers, but the level of skill as well," she said. The coaching staff of Peace Chapel relayed yesterday, "We are happy that we won the Boys division and that our boys listened to the advice we gave them and performed very well. Our A team and B team battled in the final, and we told both teams that no matter who wins we are proud that they've made it to the final."

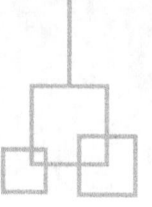

Catholic Church

(Ekalesia Katoliko)

M arist Catholic Representatives arrived in 1845 and in 1848, two Catholic priests arrived in Samoa accompanied by two Samoans. Catholic priests arrived on Tutuila to be resident priests in 1867. In 1920, the Catholic Marist Brothers introduced rugby to Samoa. A good source of information on the history of this church in Samoa can be found in the *"A History of the Roman Catholic church in Samoa: 1845-1945* by Joseph Heslin (1995), with major contributions by the late Cardinal Pio Taofinuu,

The local Catholic Church has served our local community well for many years through its *Mapuifagalele* and Hope House homes for the elderlies and the sick. These are exceptional social projects that parallels the Samoan culture which emphasize care and respect for the elderlies. Additionally, the Church's several public schools has provided education to hundreds of children in these islands for the past decades. The church's continuous support for local soldiers in the US Armed Forces has been ongoing for several years through the *Toa o Samoa* church services, held in various denominations' churches.

The Catholic church, including its auxiliary establishments, would see new changes within their communities during this decade. In 2022, a new document The Advocate. Wiggins C. 10/27/22) was released which indicated that the church is aware of some outdated teachings which include considering a more significant role for women and LBGTQ people, and other issues including sexism, clergy abuse, children of priests and homosexuality. The church however has generally continued to practice its important traditional rituals like the seven sacraments.

During this period, Pope Francis formally allowed women to serve as readers at liturgies and as altar servers, in a decree called "Spiritus Domini", that made clear that these roles are separate from the all-male priesthood. This was a formal acknowledgement in cannon law of what has already been happening in many churches around the world where women serve in these lay roles. However, by introducing the change in the Code of Canon Law, it would be impossible for conservative clerics to block women from having those roles and it was necessary as a response to the changing world. Related to this issue is the observation that America's nun population was in steep decline. According to a 2018 and 2019 Pew Research Survey, the number of Americans who identify as Christians has dropped 12% over the past decade and the group describing themselves as Catholic also shrunk leaving a crisis in the Catholic sisterhood.

Two Catholic sisters celebrated the 75th. anniversary of their Religious Profession in September, 2021, since they served as nuns for the Missionary Sisters of the Society of Mary (SMSM) from 1924-2021. They are Sr. Mary Makalita Leauai and Sr. Mary Tolotea Tauaifaga. The two Sisters also received their Apostolic Blessing Certificates from The Holy Father – Pope Francis- which were read and presented by Fr. Lui Sanele. Following the Mass, the celebration ended with a reception and entertainment provided by the families of the two sisters from Leauva'a and Manono villages.

According to the Associated Press (Sept. 18, 2021) Pope Francis urged European bishops on Saturday to listen to survivors of clergy sexual abuse and consider them partners in reform, warning that their failure to do so risks the very future of the Catholic Church. Francis issued a video message to Central and Eastern European bishops who are gathering in Poland for a four-day child protection conference organized by the bishops' conference and the Vatican's child protection advisory commission. "Only by confronting the truth of this cruel behavior and humbly seeking forgiveness from victims and survivors can the church find the way to once again be considered and trusted as a place of welcome and security for those who need it," Francis said. He also said bishops in particular must be the first to listen to victims, not the last, and must be at their service "seeing them as companions and protagonists of a common future.". There have been numerous cases of sexual abuse over the years. For example, more than 200,000 children are estimated to have suffered sexual abuse from Spain's Catholic clergy, an independent commission has found (Katherine Armstrong. BBC News) but it should also be noted that several priests have been acquitted of related allegations. For example, a jury recently found a Catholic priest in Tennessee not guilty of sexual battery against a woman who was a church member.

Father Michael Maselino Kolio, 31, was ordained as a priest during this period at the Immaculate Conception Mulivai Cathedral, becoming the first in his family to take on that path. The ordination was attended by the Archbishop of the Catholic Church in Samoa, Alapati Lui Mataeliga, priests and deacons of the Archdiocese of Samoa and family and friends of Fr. Michael Maselino Kolio. Fr. Michael is a former student of the Channel College and later studied as a seminarian in Fiji for seven years, the Philippines for over a year and in Papua New Guinea for two years before completing his studies in Samoa as part of his theological training. Fr. Michael was then assigned to the Catholic parish of Safotu.

A thousand members of the Catholic youth of Samoa gathered at Leauvaa during this period to commemorate the "Week of the Youth of Samoa" in a bid to deepen their faith and spiritual life. The program's theme was "Youth and their calling".

In 2020, a significant church congress planned for Samoa was postponed for the following year due to COVID-19 restrictions. The 5th. World Apostolic Congress on Mercy or WACOM, of the

Catholic Church was scheduled for about a week from August 10 -15 in Samoa. The postponement was re-scheduled by the Pontifical Council for the Promotion of New Evangelization at the Vatican that was to be hosted by the Archdiocese of Samoa-Apia. This is a triennial event organized "to promote the culture of mercy in the church and all over the world". Other local church annual conferences like those of the CCCS and the Methodist Church were also postponed due to the COVID-19 restrictions.

A prominent Catholic Priest, Fr. Muliau Stowers, passed away in February 2022. Stowers was a senior church member of the Archdiocese of Samoa-Apia and was the Episcopal Vicar for the Vicariate of St. Michael the Archangel in Leauvaa. The popular cleric used to appear on the Upumana television network with his regular teachings of the faith.

Satellite churches don't always follow policies/process promulgated by their mother church. For example a Catholic diocese in Michigan has been thrust into the national spotlight after its guidance on transgender members and those in same-sex relationships was shared on social media by a prominent priest and author. The now-viral guidance, issued by the Diocese of Marquette in July, states that these congregants are prohibited from being baptized or receiving communion unless they have "repented.". One advocate called it the "most egregious" guidance ever issued by a diocese. It directs the church's priests on how to develop a pastoral relationship with "persons with same-sex attraction" and "persons with gender dysphoria," and "lead them step-by-step closer to Jesus Christ in a manner that is consistent with the Church's teaching." The Roman Catholic Church has long held that **being gay isn't a sin, but being in a gay relationship or having gay sex is.** The Vatican also ruled in March that priests cannot bless same-sex unions. Controversies like this have also resulted in serious condemnation by the church. For example, in 2018 priest Juan Carlos Gavancho was punished by the church after preaching about sex scandals in the church.

In 2021, the Catholic Church's Vailima parish celebrated the sacrament of Confirmation Day on Sunday at the official home of the Carmelite Sisters. The sacrament of confirmation in the Catholic church is often held on Pentecost Sunday when Christians celebrate the descent of the Hholy Spirit upon the apostles. It's called confirmation because of the faith given in baptism getting confirmed and strengthened. More than 20 people received the sacrament on Sunday as families and friends bared witness to the occasion as they became recipients of the sacrament.

During this period, the Catholic faithful on the big island of Savai'i was able to receive the sacrament at *Logoipulotu* where more than 200 people gathered. The sacrament can only be conferred by the archbishop, and he was the one who led the celebration on Sunday at the *Vailima* parish church. This sacrament can only be given to someone who has accepted his or her Catholic beliefs wholeheartedly and specifically those who are 14 years of age and above.

In early 2022, many Catholic faithful visited a family resident in Leufisa, Samoa to witness what some called the "weeping Jesus Christ statue". The statue is inside the home of Catholic layman Brother Vincent Dominica Rozari of the Order of the Divine Mercy. The *Samoa Observer* took photos and Brother Vincent collected an "oil-type" liquid substance from the foot of the statue. In the US and other countries, several of these types of events have happened for several years and some of the tears and blood from these statues were scientifically analyzed. After more than ten years and using some of the top laboratories in the world, some of the analysis of blood samples from these statues were found to be from a woman but the analysis was unable to extract DNA from the samples!. I think this will be a very complex and prolonged issue because the issue can be related to Exodus 20:4.

In Samoa during this period, native lady *Toaipuapuaga* "Toa" Opapo (mentioned in another section and also in my previous book) brought home a small bottle with some holy liquid, allegedly tears from the Virgin Mary's statue in her church. She used the liquid several times to provide some comfort and healing for her family. Every time the liquid seems to run out, Toa would find it be miraculously re-filled (TV4 Catholic Samoa "*O Lou Valaauina*). Overseas readers can contact me if they want to get directions to her home. A wonderful lady. A stigmata phenomenon also occurred in Upolu, also happened to this lady, where she showed visible bleedings on her feet, hands and forehead around Easter. Stigmata is defined as "a mental or physical mark that is characteristic of a defect or disease". The bloody marks on Toa's body[7] corresponded well with Christ's wounds on the cross. Furthermore, Toa provided several messages that were later confirmed by Malua Theological College (MTU) instructors and the Israel office in New Zealand as genuinely written in Hebrew. Toa later travelled to the Vatican where the Pope held that the messages were from God. Apparently, there were no messages pertaining to the Prosperity Gospel preached in Mega Churches. Since I believe these messages were from God after careful scrutiny by Bible College Instructors, a few Israel experts, the Pope and several Catholic Priests, I strongly believe the Prosperity Gospel would have been mentioned by God in these messages since these were modern-day messages for present day worshipers. Note: Readers who want a photo of this stigmata event can contact me for a copy.

American Samoa was in Code Red of the COVID-19 restrictions during the 2022 Ash Wednesday and Bishop Brown announced that Mass was to be streamed on the Samoa Pago Pago Facebook. Families were to have their own distribution of ashes and should burn the palm leaves from the previous year's Palm Sunday or use ash from any burnt branch. Catholics were encouraged to gather and explain the meaning of Lent, traditions of prayer, penance and fasting.

When Russia invaded Ukraine during this period, the Archbishop of the Catholic Church in Samoa appealed to Catholics to pray for

---

[7]

peace during the Ash Wednesday Mass held at the Immaculate Conception Cathedral at Mulivai. Foreheads were marked by the priests and deacons. The archbishop reminded the congregation that there were a lot of Christians in Russia and in the Ukraine. I'm not going to pray for wicked people who killed women and children and bombing churches and hospitals. According to Wall Street Journal correspondent Mathew Luxmoore, the Russian Orthodox Patriarch Kirill has endorsed Putin's war in Ukraine[8]. If this was the only church on Earth, I would ask God for forgiveness, but I would not attend this church.

In 2021, hundreds of Catholic members from the Archdiocese of Samoa gathered at the Tofāmamao Pastoral Centre, Leauva'a from Sunday 5th to Wednesday 8th September to celebrate a spiritual renewal of the Legion of Mary. This conference consists of various church groups who honor the Blessed Virgin Mary as the Mother of the World. One of the groups, the Legion of Mary celebrated its 100 anniversary on Tuesday 07 September which was founded by an Irish layman and civil servant, Fran Duff on 7 September 1021. The Legion of Mary is a voluntary international association of members of the church through prayer, visitation to the sick and poor, attending Mass and learning more about the Catholic faith. Hundreds of members of the Legion of Mary and Catholic members from close to ten Vicariates of the archdiocese of Samoa took part in the evening celebration that commenced with a procession that followed the Statue of Mary being carried from the Tofāmamao Centre to the cave at St Therese. The candle lit service was followed by cultural performances by the various Vicariates. One of the activities for the Legion of Mary is to pay visits to those in need financially and in health. Each Vicariate concluded their performances with monetary offerings of more than $1,000 each to help support the hosting of the conference.

In March 2022, the American Samoa Catholic Church invited the public to join them in their "Dedication of Ukraine and Russia

---

8 The Daily Digest, zeleb.es 1-17-23

to the Immaculate Heart of Mary" online Mass scheduled for March 25, 2022 and streamed from Fatuoaiga on the Diocesan Facebook. According to Bishop Brown, the Consecration Prayer was to made available online for everyone to read together.

During the 2022 Easter weekend, Bishop Peter Brown reminded the public that "the eternal message of Easter is hope, Jesus Christ died and rose again that we might see light in the midst of darkness". He added "The world around us appears clothed in darkness with pandemics (COVID-19) and war (e.g. Russia invasion of Ukraine) bringing suffering and pain to many.

In early June 2022, a 14-year-old Jon Bosco of Leauvaa Uta informed Archbishop Alapati Lui Mataeliga of a message from Mother Mary that was to be told to the whole country and that it was up to the church to do the work required. The work was to begin on May 14 from 6am to 3pm where priests were to conduct Mass and pray the rosary. Three months after the first vision, an oil-like liquid substance continues to seep from the statute. Many Catholic faithful brought their own statutes of Mary and Jesus to Jon's home and left with vials of oil taken from the weeping Mother Mary (Samoa Observer, 5/13/22).

Deacon Kasiano Leaupepe passed away at the age of 86 in 2022. He was a former Chairman of Samoa's National Council of Churches. He was an eloquent orator and was not afraid of admonishing church and political leaders. He had a good sense of humor and noted that a lot of Catholic priests have traditional tattoos. (a few Congregational and AOG pastors also have the *tatau*). He also led a NCC delegation to seek a peaceful outcome to the HRPP elected members being locked out of Parliament after the elections.

In August 2022, Chancellor and Parish Priest of Holy Family Parish Tafuna, Father Kolio Etuale, was chosen to be the Coadjutor Bishop of the Diocese. This was according to Bishop Brown, Head of the Catholic Church in American Samoa. The announcement was also made by Pope Francis on Vatican Radio in Rome. Father Etuale has a BA and Masters Degrees and was ordained priest in American Samoa in May 2003. The consecration or the Episcopal Ordination, was planned for November 4, 2022 for the Feast of Str Charles Borromeo. The Governor of American Samoa later issued a public congratulatory message and blessings to Father Etuale. The 49-year-old Father Etuale, would become the first Samoan Catholic Bishop appointed for American Samoa. Foreign dignitaries and relatives from Australia, New Zealand, Hawaii, US and Samoa attended the celebration. Traditional presentations like the *ava* ceremony and *ta'alolo* were part of the program.

In American Samoa, the Cathedral Parish of the Holy Family at Fatuoaiga celebrated the completion of a major renovation of the cathedral during a combine Mass led by Bishop Peter Brown. The project included a new ceiling, a new dome and roof, various repairs of the pews and the general interior and exterior. Part of the funds of just over a million dollars were from payments for the use of the Fatuoaiga Pastoral and Culture Center as a quarantine center during the COVID-19 pandemic. The renovations were completed just in time for the ordination of Bishop Fr. Kolio Tumanuvao Etuale.

In very rare occasions, the confessionals endorsed by the church are not so confidential. According to author Nylee Ausier (Yahoo 12-22-22, r/facepalm) a women told readers that she blamed a priest for ruining her marriage after he told her husband she confessed to cheating.

Bishop Peter Browns 2022 Christmas and New Year's message included reflection on those that have passed away but also remined the community of the joy and hope of life.

Two dreadful events marred the end of 2022 for the Catholic Church. First, the 95-year-old Pope Emeritus Benedict XVI passed away at his home in the Vatican. Thousands of people gathered at the St. Peters' Square and his body lie in state in St. Peter's Basilica for three days. The funeral was presided over by Pope Francis, an unprecedented event in the modern Catholic Church. There were, however, missing elements that pertained to a reigning pontiff such as the supplication of the diocese of Rome and the Eastern Churches and the Final Supplications. Pope Benedict resigned in February 2013 at the age of 85.

Secondly, the 95-year-old Bishop Quinn Wetzel, who led the Diocese of Samoa Pago Pago for 29 years passed away on December 30, 2022. Bishop Quinn passed away at his residence at Maryknoll New York. He was a Maryknoll Priest for 67 years and a Bishop for 36 years. This will be an unprecedented chapter in the history of this church with a reigning Pope eulogizing a retired one. It was announced during this time that Bishop Peter Brown, head of the Catholic Church in American Samoa was to preside at the Requiem Mass and the Celebration of Life for this first Bishop of the Diocese of Samoa-Pago Pago. A memorial Mass was also planned for the Holy Family Cathedral, Fatuoaiga on the Martin Luther King holiday the following week.

The Carmelite Sisters at Vailima received some special visitors towards the end of 2022. The Toa Samoa Rugby League Team visited the Nuns and were hosted by the Vailima Catholic parish. The Carmelite is the order of Catholic nuns that initiated the monthly Toa o Samoa church services in American Samoa, asking God's protection of Samoan soldiers in the US military. Archbishop, Monsignor Alapati Lui Mataeliga also invited the Manu Samoa 7s team around this time to celebrate his 70th. Birthday as well as his 20th. Episcopal anniversary as the Archbishop of the Catholic Archdiocese of Samoa and also 45 years since his priesthood ordination. The archbishop was born on January 4, 1953 in Sataua, Savaii. He was ordained for the presbyterate of the Diocese of Samoa and Tokelau in 1977. Acknowledging these patriotic achievements by this church seem to be in contrast with the Jehovah's Witness convictions.

In early 2023, Bishop Brown urged bishops of Oceania to reflect on their place in the church and to act on Climate Change. This was during a conference for the Catholic Bishops of Oceania (CEPAC) in Fiji. He said that he was told stories from people who have been displaced by rising sea levels and hope the church will champion the climate change cause. One way to accomplish this is for every Catholic parishioner to campaign hard and vote for Al Gore to become US President.

Funeral services for the longest serving priest in the Catholic church in both Samoas were held during this period. Services for Monsignor Etuale Lealofi were held at the Holy Family Cathedral Fatuoaiga. He was 82. He was ordained a priest in Apia in 1969 and was one of the first priests in the South Pacific to become a canon lawyer, a lawyer for the church. He was Rector at the Moamoa Theological College and also the Pacific Regional Seminary in Fiji. After the Diocese of Samoa-Pago Pago was established in early 1980's, he was one of the core staff of the infant diocese and was also a key staffer at the Fatuoaiga Pastoral and Cultural Center where he was a lecturer for diaconate formation and also planned the first diocesan synod in 1990. He was been cared for at the Hope House at Fatuoaiga. Church pioneers are leaving for greener spiritual pastures.

Towards the end of national celebrations of Samoa's 60th. Independent anniversary, the Catholic church at Lealatele offered to host an event to celebrate the 178 years anniversary of the arrival of the Catholic church in Samoa. The village believed that the Catholic church first arrived at their village 178 years ago and they are proud to host government officials and members of the public in May 26-27, 2023 to commemorate this event.

In early 2023, the United States Conference of Catholic Bishops (USCCB) published a letter (Nerozzi, Timothy. 3/1/23) warning US Senate members against supporting the revival of the Equal Rights Amendment. USCCB voiced concern regarding a motion in the Senate that would remove the old deadline for ratification of the Equal Rights Amendments – a proposed amendments to the constitution that would guarantee "equal rights under the law" to all Americans regardless of sex. The USCCB warned that modern interpretations of the decades-old legislation would expand far beyond the original intentions of the bill and force states to make abortion-related accommodations. A consequence of this amendment would likely be federal funding for abortions. There have been cases where these abortion laws have resulted in the deaths of women and babys and I had advocated that abortion is a personal choice and therefore **government should stay out of people's bedrooms**. Additionally, any new laws relating to abortion should have a clause affirming that the legislative sponsors of such bills are personally responsible for any and all expenses related to abortion casualties. In late 2023 a women flee Texas for an abortion when both her and her unborn baby's medical condition were in danger. There should be a law in Texas requiring the lawmakers who passed related laws to personally pay for the woman's expenses. Personal choice in abortion is a **women's right and is far more appropriate than any religious beliefs**. If God gave man, beginning in Genesis, free will to make decisions then who are we (public and government) to make personal decisions affecting a women's life and body?. Abortion associated beliefs have been debated for decades and will continue to be controversial until Jesus returns to provide the correct answer since it's a matter of interpretation for now. No human can correctly interpret Biblical concepts because it entails reading the mind of God or actually talking with God and asking HIM for explanations and the correct interpretations. Let me know of your thoughts.

In American Samoa, students of the Catholic Church decorated floats, gave speeches, did Tik Tok dances and choral reading at the Fatuoaiga during the March 2023 celebration of the *Laudato Si,* which is an encyclical, or letter from Pope Francis calling on everyone to take care of our home earth. The performances underscored the concepts of pollution, recycling, natural resources and the environment. Bishop Peter Brown blessed the *Laudato Si* publication and copies were given to guests which included the Governor, priests, catechists and several government officials. Former Samoa Head of State, Tuiatua Tupua Tamasese later met with officials of the Dicastery for Promoting Integral Development in Rome to continue discussions on this matter. There were also discussions of a planned the Pulemelei Research Center and Biosphere Reserve to further the movement for ecological restoration.

April 22, 2023 was Earth Day and all parishes of the Catholic Church in American Samoa planned to clean up their churches, homes and villages. The Diocesan-wide Clean Up Day fell within the recent launching of the *Laudato Si*.

During the American Samoa Flag Day of 2023, the Government of American Samoa offered a fond farewell to Bishop Brown as he was retiring at the end of the month. The Diocese *taualuga* for the occasion led by the bishop netted around $9k.

The leader of the Catholic Church, His Grace Archbishop Alapati Lui Mataeliga passed away in New Zealand in 2023. He was 70. Bishop Brown, priests, deacons, Catechists and people of the Diocese of Samoa-Pago Pago immediately offered their prayers and sympathy to Samoa. Alapati, at 24, was believed to be the youngest Samoan seminarian ever to be ordained a priest in 1977. He was appointed by the Vatican to be Archbishop on November, 16, 2022. The archbishop also served in American Samoa at Laulii, Fagatogo and the Holy Family Parish at Fatuoaiga. He was vocal and a constant critic of HRPP and its leader. In 2009, he led a large group of youths to the World Youth Conference in Australia that was attended by Pope Benedict. The motorcade that brought his coffin from the airport to Apia was very impressive, with hundreds of students in uniforms, Chiefs in their traditional wear, members of the clergy and the general public lining the main road. This was a very sad and moving day for all members of the Catholic Church. Special services including a mass by the Catholic Youths were also held. *Ia manuia lau malaga*.

In April 2023, the Most Reverend Bishop Hugh Brown officially stepped down as the Bishop of the Samoa-Pago Pago Pule'aga with the installation of 49-year-old Bishop Kolio Etuale. Etuale was appointed as coadjutor bishop on August 4, 2022 and was consecrated by Bishop Brown in November 4, 2022. The celebration acknowledged the work of Bishop Brown who mentioned the feast of Catherine of Sienna, a 14th. Century saint, which is celebrated on

April 29 – the day of Etuale's installation. The program included the presentation of a Samoan *to'oto'o*, *fue* and the Coat of Arms. Bishop Etuale is the first Samoan priest to be chosen bishop for the Catholic Church in American Samoa and was ordained priest in American Samoa in May 29, 2003.

The 2023 Independence celebrations in Samoa included a visit by government officials to Savaii for a service commemorating the arrival of the Catholic Church in Samoa.

The 5th. World Apostolic Congress on Mercy (WACOM) started its events in Samoa, in May 2023. The theme of the 4-day event was "Ocean of Love that Envelopes the Whole World". This was the first time this prestigious conference has been held in Oceania. The first event was held in Rome in 2008. Samoa was host to about 500 overseas delegates who travelled from as far as Asia, Europe and from around the pacific. Local church leaders invited the public to the commemoration of the 178 years of Catholic Church in Samoa and also mentioned a tribute to the late Archbishop Alapati Lui Mataeliga. Archbishop Franz Pewter represented the Pope. The issues for the conference included: Message and Mission according to St. Faustina; revisiting the Spirit of the Extraordinary Jubilee of Mercy, 2017; The face of Christ is the Face of the New Evangelization – Do we see Christ in the Poor and the Migrants of our times?.

A 12-member delegation from American Samoa attended the World Youth Day of the Catholic Church held in Lisbon, Portugal during this period. This event brings together Catholic Youth from around the world for faith renewal, spiritual enrichment, fellowship and cultural exchange. The event was also an opportunity for the youth to meet with the Pope and receive an invitation to consider a religious vocation.

In American Samoa, the International Community of the Cathedral of the Holy Family Catholic Church held its carnival at the Fatu-O-Aiga Community Hall. The event was opened to the public and included face paintings, games, snacks, prizes, various fun activities and family fellowships.

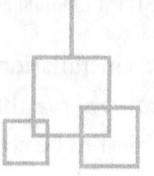

Church of Jesus Christ of Latter-Day Saints

(Ekalesia a Iesu Keriso o Aso o le Toe Afio Mai)

I n 1863, Missionaries Kimo Pelia and Samuela Manoa established the Church of Jesus Christ of Latter-Day Saints on Tutuila and the church was later formally organized in 1888 when Joseph and Florence Dean arrived. In 1892 the church started chapel schools. In 1903, the translation of the "Book of Mormon: Another Testament of Jesus Christ" was completed. In 1890, there were about 30 members in the Samoan islands. The church started Chapel Schools in 1892. On eBay, one can find out how to purchase a stamp commemorating the arrival of this church in Samoa. The information reads: Samoa 1988 Centenary of the LDS" and the following words appear on the stamp: *Great are the promises of the Lord unto those who are upon the isles of the sea"*

As of 2021, it was estimated that there are about 16,000 LDS church members in American Samoa attending 43 congregations on Tutuila and about 100,000 members with 205 congregations in all the Samoan Islands. It is believed that the Samoan Islands has the second most LDS church members per capita in the world, behind Tonga, and also has the largest body of LDS members in Oceania outside New Zealand and Australia.

According to the Star Insider (1-12-23), Mormonism Church was founded by Joseph Smith in the 1820s and follow the religious traditions of the Latter-Day Saints (LDS). This church and the Church of Scientology are believed to be the only religions founded in the US. LDS believe that the God of Abraham was a person living on a distant planet that circled a star called Kolob and that all good religious believers become gods when they die; believe that the garden of Eden was in Missouri; that women who are not believers of this church become a harem to a Mormon man in the afterlife; members are required to tithe; parishioners have "temple garments" which consist of white T-shirts and knee-length boxer shorts; on Sundays, members are not allowed to watch television or films, and cannot play sports; if any child under eight years old dies, they automatically go to heaven; members are not permitted to consume anything that pollutes the body or affects the mind (e.g. alcohol, tea and coffee). Jake, a professional musician friend of mine and a Mormon, should be penalize by this church because he drinks beer nearly every day including Sunday. An interesting belief that is related to the Samoan culture is that its members are prohibited from getting tattoos. Many Samoan women have *malu* and males have *tatau* or pe'a tattoos including some pastors. In the field enforcement, investigators in many countries have welcomed the presence of tattoos on the victims' bodies as these facilitate their identifications.

In mid-March 2020, The Church of Jesus Christ of Latter-day Saints began to fly hundreds of their missionaries back to their home countries due to COVID-19. Limited by the pandemic, The Church of Jesus Christ of Latter-day Saints scaled back physical outreach and intensified virtual proselytizing, calling on the flexibility and social media savvy of young missionaries.

In late September 2021, Members of this church prepared for its 191 General Conference which is a 2-day event scheduled for broadcast across Samoa. These conferences are held to "commend true merit; express gratitude for divine guidance; to give instructions in principles, in doctrine, in the law of gospel; to proclaim the restoration and to admonish and inspire to continue in greater activity" according to President David O. McKay. The meeting was to originate from the Conference Center in Salt Lake City Utah and consists of five sessions. According to the church, the conference is translated into 194 languages and is viewed by members in 221 countries. The total church membership at this time in Samoa was 83,740 in 20 stakes and 162 congregations. General Conferences take place every six months and the October 2-3, 2021 conference is digital only.

In 2021, According to the Associated Press, the Church of Jesus Christ of Latter-day Saints announced that masks will be required inside temples to limit the spread of COVID-19. Church leaders said in a statement that masks will be required temporarily in an effort to keep temples open. The message was the latest in a series of statements from church leaders encouraging masking and vaccination efforts against COVID-19.

The official ground breaking ceremony for the construction of the Pago Pago, American Samoa Temple on the Ottoville Road was held in late October 2021. It was reported that the LDS membership has grown to more than 16,000 and that the church's two Hawaiian missionaries who brought the church to the Samoan Island arrived at Aunu'u around 1862 (*Talanei, Oct 2021*). Building a Temple in Pago Pago to serve both American Samoa and Samoa was first announced in 1977 but later changed to building the Temple in Apia to serve members of the Samoan islands, according to church leaders and information shared with journalists during a pre-ground breaking news conference held at the Church's Central Stake Center at Ottoville, Tafuna.

Mormon Church
Caption: New LDS Temple in Tafuna

A few members of the American Samoa Legislature met with Church Elders O. Vince Haleck, K. Brett Nattress and Fa'apito Auapaau at the Senate in November 2021 to present a Senate Concurrent Resolution to Nattress on behalf of The Church for the significant contributions that the Church has donated to American Samoa. Nattress expressed gratitude and love to the leaders and all the people of American Samoa for their efforts to raise strong families, love God and their fellow brothers and sisters, and to do good in their homes and communities. In the resolution, it commends and congratulates LDS on the recent ground breaking of the American Samoa Temple, and acknowledged the presence of Nattress, who presided over the ceremony. The Legislature also offered sincere gratitude of the territory for all of LDS's generous humanitarian efforts for the people of American Samoa. Ground breaking for the 17,000 square-foot, single-story Temple was held October 30th and the project will include housing for the Temple president, matron and Temple missionaries as well as a distribution center. The non-binding resolution notes that The Church has played a large role in American Samoa through its numerous humanitarian efforts regardless of race, religious affiliations, and private and public organizations, and the territory is indebted to the Church for the many blessings it bestows upon the territory for many years. The resolution explained that LDS is credited for its great investments made in LBJ Medical Center having contributed immensely over the years. For example, in 2013, The Church presented a digital mammography and biopsy machine to the hospital. In 2019, "members of a dynamic team serving as welfare and self-reliance missionaries" were called to the territory to train and provide support for LBJ's medical professionals. Elder Dr. John Edwards and his wife Becky, Elder Dr. Greg Patch and his wife Janene and Elder Dr. Robert Keddington were on assignment to LBJ for 18 months. Elder Dr. Edwards, worked specially to set up the first ever knee replacement clinic in the territory and performed over 100 successful surgeries. During recent natural disasters, the Church opened its doors to families across the territory that were temporarily displaced, according to the resolution, which also declared that The Church's Relief Society women's organization has over many occasions supplied clothing, blankets, and pillows to families in need. Furthermore, The Church through the "LDS Charities" made a contribution of $25,000 to the Pala Lagoon swimming pool center. They have also donated commercial washing machines and vital signs machines to Hope House and also donated paint to various schools on island and continue to perform community service through efforts to clean up Lion's Park.

LDS presented to the Governor of American Samoa digital copies of records that date from 1900 to 1974 that were copied from microfilms — which were damaged and could only be saved by digitization to a computer format. These are records of valuable history that would have been permanently lost were it not for this work by the Mormon Church. Historical records are vital to every culture and according to the Center for Samoan Studies Director "We seem to lost information...we have modern technology at our disposal to go one step further and to begin recording our own stories". At the meantime, the American Samoa Office of Vital Statistics have lost many original birth certificates of several US Nationals that came by our office (American Samoa Legal Aid) for legal assistance.

Aunu'u island has a special place in the history of the Church of Jesus Christ of Latter-Day Saints. It's where missionaries baptized the first church members in the Samoan Islands in 1866.

During this period, members of the Church on Aunu'u were delighted to meet with Elder K. Brett Nattress, from the Pacific Area Presidency, accompanied by Elder Faapito Auapa'au, the Area Seventy for American Samoa, and President Sonny Aiono. Elder Nattress was on island to attend the groundbreaking of the LDS Temple in Ottoville and said that he was deeply touched by the experience of meeting the wonderful Saints in Aunu'u, and that their faith in Jesus Christ and devoted discipleship were inspiring. Lotoa Lotoa Jr. Bishop of the Aunuu Ward greeted the visitors and said that it was a great experience and blessing to have the Church leaders visit their small Island. The party walked around the entire island before returning to the small chapel for a traditional Samoan lunch of traditional food including fish and taro. Several pioneer members recounted the history of the island and the Church and one such story "The Miracle of the Sand" described the construction of the chapel where the people who went to build the chapel were astonished when they found a huge pile of sand at the construction location. The only source of sand, which was critical for the construction of the new chapel, was the deep surrounding sea and it would be a great expense to get it. A local miracle I guess.

Jeff Green, a billionaire thought to be the richest person from Utah, resigned from the Church of Jesus Christ of Latter-day Saints, writing in a letter to the church's president that he believes the institution has "hindered global progress in women's rights, civil rights, racial equality, LGBTQ rights," (Salt Lake Tribune). Green, the chairman and CEO of The Trade Desk, a technology marketing company, informally left the Mormon church a decade ago. But in a letter to church President Russell Nelson, he officially resigned and requested the removal of his records. The Tribune also reported, Neither Green nor the church have responded to requests for comment about the letter. Green said in his letter that most of the church's members are "good people trying to do right" but that he believes "the church is actively and currently doing harm in the world." "The church leadership is not honest about its history, its finances, and its advocacy," he wrote, according to The Tribune. Green wrote that he will donate $600,000 to the LGBTQ advocacy group Equality Utah as the first major donation from his family foundation Data Philanthropy. According to The Tribune, he added that almost half of the money will go to a new scholarship program for LGBTQ students in Utah, including those who "may need or want to leave" his alma mater, Brigham Young University, which is sponsored by the church. Executive Director Troy Williams said Equality Utah is "incredibly grateful for Jeff's generosity and support." About 62 percent of Utah residents and about 86 percent of the state's lawmakers are members of the LDS church, which opposes same-sex marriage and more recently the Equality Act, a federal bill that would protect LGBTQ people from discrimination in housing, employment and jury service, among other areas of life.

Similar instances have happened in the Samoan Islands where some people who have been members of the LMS church have left to join some Pentecostal church. Their main reason was "this is the first time they experience the presence of God's Spirit" when they joined Pentecostal churches. The LMS Missionaries have a different approach or understanding on this subject. Another reason people leave is the hefty monetary donations required of them from congregational churches.

During this period, LDS donated tables, 110 chairs and boxes of books to Auala Primary School in Samoa. The school facilities were also improved with funds from the Government of Japan.

In October 2018, a few of Utah's top lawmakers, representatives from The Church of Jesus Christ of Latter-day Saints and medical marijuana advocates met at the State Capitol. They were joined by the Utah Medical Association and law enforcement groups to announce a deal to legalize medical marijuana across the state. Everyone was there in large part because the church had decided they would be. Despite the national popularity of legalizing marijuana, 14 states — mostly in the Deep South and the Great Plains — have not embraced it for medical use. The Mormon church helped greenlight a medical marijuana program, industry advocates may use to succeed in deeply conservative places like Idaho or Kansas.

Season 3 of the Book of Mormon video series in the Samoan language was launched by The Church of Jesus Christ of Latter-day Saints during this period. The President of the Apia Samoa Central Stake announced Season 3 of *"Vitio o le Tusi a Mamona"* in a Facebook post. The first episodes of the Book of Mormon Video Series in Pacific languages were launched about three years ago in September of 2019. This is one of the churches' initiatives as it embraces new technology like the use of iPads by its missionaries during their proselytization visits to peoples' homes.

According to the Associated Press (Brady McCombs, 6-14-21) Leaders of The Church of Jesus Christ of Latter-day Saints (LDS) with a 16.5 million membership, announced $9.25 million in new educational and humanitarian projects during this period as they seek to build on an alliance formed with the NAACP in 2018 as part of the Faith's efforts to improve race relations. LDS will donate $3 million to fund scholarships for three years for Black students through the United Negro College Fund and give $250,000 to create a fellowship for students from the United States to travel to Ghana to learn about slavery. LDS will also give $6 million to fund three years of humanitarian aid aimed at helping underprivileged people in six metro areas of the United States. LDS announced the initiatives ahead of Juneteenth, a holiday celebrated on June 19 that commemorates the date of the emancipation of enslaved people in the United States. The church's partnership with the Civil Rights organization is part of the religion's efforts to improve race relations. Until 1978, the church had a ban on Black men in the faith's lay priesthood that was rooted in the belief that black skin was a curse. The issue lingers as one of the most sensitive topics in the religion's history.

The LDS Pesega campus was recognized by the Government of Samoa for its example of beauty and cleanliness in an awards ceremony in 2022. The ceremony was part of the country's "Beautiful Samoa" campaign which aimed at staving off illness and diseases and to protect the nation's natural environment. LDS received a certificate and a donation from this campaign.

LDS church presented a 40-foot container filled with walking aids, medical supplies, wheelchairs, boxes of face masks and prosthetic kits to the Samoa Government in July 2022. This was one of the on-going activities by the partnership between the church and the government. The container was presented by the Welfare

& Self-Reliance Manager and these types of donations had been on-going since 2014.

In late 2022, a member of this church, Mitzi Semo, was appointed to the General Relief Society Advisory Board, which assists the Relief Society General Presidency of the church. She was a former Hawaiian Airline Station Manager and her husband served as Bishop of the Mesepa International Ward. Mitzi is from the Orem 12st. Ward (Samoan), Provo Utah 1st. Stake (Tongan). She also served in many other capacities including full-time missionary in Paraguay Asuncion mission.

The Vailima Ward and other ward teams held a contest for the Apia Samoa Central Stake Athletics Day 2022 at the Apia Park in October 2022. More than 700 members of the church from ten wards competed in the spirit of fun and fitness. This athletics day was part of the stake's Family Proclamation aimed at raising awareness on the fight against violence and to promote healthy living, according to the President of Samoa Apia Central Stake. The President added that the event also underscored the importance of the family unit in God's Plan of Happiness. The event also promoted family health and fitness as statistics showed increases in the number of local people with heart problems, diabetes, high blood pressure, lung diseases gout.

The church released a statement in late 2022 supporting of the US Senates Respect for Marriage Act. In the past, the church wrote that acting on same-sex attractions is a sin and had excluded the children of same-sex couple from baptisms and naming ceremonies.[9]

In late 2022, this church donated a 15-seater Hiace van to the *Mapuifagalele*, home for the elderlies. The donation was accepted by the Little Sisters of the Poor and presented by the Leader of the Apia Samoa Stake of LDS. The Little Sisters of the Poor had requested LDS for a vehicle to be used for various tasks of the organization.

The Church group from Aleipata delivered various donations

---

[9] JEZEBEL. Cheung, Kylie 11-16-22

for the Tanumalala Prison in early 2023. The Upolu East Stake comprised of around 3,500 members and is one of twenty in Samoa and five in American Samoa. A stake is a Mormon term for a group of congregations, called Wards - in a geographical area, similar to a diocese in the Catholic church. During this time church membership was estimated at 80,437 in Samoa.

Miss Samoa, Haylani Pearl Kuruppu attended the Apia Samoa Central Stake Youth Conference in early 2023 and gave a speech encouraging the youths of 11 wards to attend church. The event was also in partnership with the Ministry of Natural Resources and Environment (MNRE) in a drive to plant one million trees to combat Climate Change. The theme of the event was: Jesus Christ is the Strength of Youth.

Gynecologist Dr. Audrey Tarr, an LDS missionary volunteer doctor ended her volunteer work at LBJ in early 2023. Dr. Tarr was one of several LDS volunteers who worked in American Samoa at the hospital. This church has provided various types of medical assistance to the local LBJ hospital in the past, for several years.

The Church announced in early 2023 that it is planning to donate roughly 20,000 acre-feet of water rights to the Great Salt Lake which has shrunk to its lowest levels ever. The church has at least 75,000 acre-feet of active water rights. The donation is about 2% of what's needed to keep the lake at its current level.

Eleven Wards from the Apia Samoa Central Stake conducted a Community Service Project for schools and villages in the Motootua, Vaimoso and Alafua areas in early 2023. The project cleaned up school compounds, public footpaths, roadside areas, drainage and ditches. According to Church officials, this is an annual project that occurs during the month of February and eleven units worked within their own geographical boundaries.

The Samoa Victims Support Group in Apia was blessed in early 2023 through a visit from the Pavaiai Third Ward Youth of American Samoa. The visiting group spent quality time with children at the Campus of Hope, sharing the Gospel and enjoying various fun-filled activities. The group was led by Bishop Tuinei and were in time to impart the message of Easter. At this time, Bishop Brown also relayed his Easter message saying that "Unless you are reborn in the Spirit you cannot see the Kingdom of God". He added: Easter is a time of rejoicing. It is a time for jubilation for we live in the light of his kingdom".

Tafuna Ward 3 held its Annual Conference for the Middle Stake in May 2023. Entertainment groups from both Samoa and American Samoa provided a variety of items.

LDS Missionary doctors who saw patients at the Manua islands in early 2023 were impressed with traditional lifestyle and good diets of the residents. The LDS team included cardiologists, radiologists, and an orthopedic surgeon. The group were performing an outreach project with the Department of Health. These kind volunteers also conducted nutritional awareness in the schools.

The Church's Vaiola High School in Savaii mourned the passing of one of their beloved school teachers who died in a car accident in early 2023.

In June 10, 2023, a Children's Book called "Faith, Hope and Miracles" was launched at the Church of Jesus Christ of Latter-day Saints meeting house on the island of Aunuu, American Samoa. The book, in English and Samoan, was produced by the Pacific Area of the LDS Church and it recounts true stories of the Church's history in the Samoan Islands. The Pacific Area Presidency noted that the first missionaries arrived in these islands in 1863. Elder K. Brett Nattress, presided at the meeting. A historical marker was also unveiled and dedicated during this event, and it commemorates the 1863 arrival of the first missionaries, Samuela Manoa and Kimo Pelio on Aunuu.

The Annual "For the Strength of Youth" (FSY) 2023 Conference of the LDS Church was held in American Samoa during this period. The youth experienced a little of what life as missionaries might be like, in addition to group praying, classes, devotionals, studying the Bible and the Book of Mormon and various activities. Elder Taniela

B. Wakolo and his wife Sister Anita Wakolo joined the youths via Zoom. The event was mainly for youths 14-18 years.

In September 2023, President M. Nelson announced a new temple for Savai'i at the closing of the General Conference of the Church. The President explained that the temple is a place of revelation. Savai'i was the first of 20 new temples locations named following the General Conference. Savaii, with a population of about 45,175, has 47 congregations and an LDS Church High School-Vaiola College. The Samoa temple in Pesega was dedicated in August 5, 1983 by Elder Gordon B. Hinckley of the Quorum of the Twelve Apostles. This temple was rebuilt after a fire in 2003 and rededicated in 2005. There are ten operating temples in the South Pacific.

President M. Russel Balard passed away during this period. He did visited these islands and one of his favorite sayings is 'Keep it Simple". His favorite scripture was John 14:15 according to his son. He often visits and bless patients in hospitals. One of his last words he said to his son was "I am Cleaned".

The LDS Church "Light of the World" organization gave various donations around 2023 Thanksgiving to the LBJ Hospital.

The Apia Samoa Central Stake celebrated its tenth anniversary in Apia during this period. The celebration started with a parade led by the Samoa Police Band. Series of activities spanned 10 days and included a Zumbata Session, an Awards Nights, Prom Night and Leadership Training sessions. Apia Samoa Central Stake is made up of 11 congregations.

Below is the photo of the partially-completed LDS Temple build at Tafuna, American Samoa, taken on DATE. I had hoped the Temple would have been completed when I publish this book so this book would be the first to publish a photo of this important church structure. I would have waited but I'm very disgusted with the emergence of Artificial Intelligence with its fake and non-natural products that I decided to publish this book ahead of the planned launching date in case I accidentally use its Deep Fake imitations and false information. False and "mechanical" materials provided by AI prompted Hollywood writers to request for protection from AI during this period.

New Mormon Temple

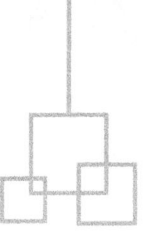

Voice of Christ Full Gospel

T he American Samoa chapter, a prophetic ministry, was founded by the late Rev. T.K. Tilo but was started in San Francisco. The late Elder Rev. Rudolf Tilo later took over followed by Ted and Rachael in January 2019. There are branches in New Zealand, Australia, Fiji, Tonga, American Samoa, the US and in Samoa. The church also has affiliated churches in Pakistan.

In American Samoa, Rev. Elder Umaga presided over the funeral services of Rev. Elder Perekina Enesi in April 2022.

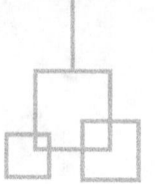

Jehovah's Witness
(Molimau a Ieova)

J ehovars Witness was founded in 1870 by Charles Taze Russell. The name Jehovah is taken from tetragrammaton, written as YHWH or JHVH and articulated either as Yahweh or Jehovah. The group was originally called the Watch Tower Society because Taze published a magazine called Zion's Watch Tower and herald of Christ's presence. This group use their own New World Translation of the Bible. They don't believe in trinity and believe Jesus is not equal with God as explained by John 14:28 "The Father is greater than I am". They do not believe the Holy Spirit is equal with the Father. They don't celebrate Easter and Christmas and try to remain politically neutral. Its members are discouraged from serving in the military (like the Jewish Ultra-Orthodox church members) and politics. They don't believe in blood transfusions even in life-or-death situations. There are reportedly 8.3 million members around the world http//www.crosswalk.com/).

During this period several Jewish students join the army to fight Hamas. Maybe Jehovar's Witness know something that these students don't.

A trial held in the US saw an attorney allegedly unmasking this group's leader's lacking knowledge of the Greek language; a requirement in understanding the Bible. According to one local pastor, this can cast doubt to the group's interpretation of the Bible since understanding this language (in addition to Hebrew and Aramaic) is essential to the interpretation of the Bible, especially for the leader of a church.

This group has made several erroneous prophecies regarding the return of Jesus Christ and the end of the world. For example, they predicted that the Time of the End started in 1799; Jesus returned invisibly in 1874; the End of the World will be in 1914, then they changed it to 1915 and later to 1918. After his numerous faulty prophecies, Russel died in 1916. The second leader, Rutherford continued with these flawed prophecies, predicting the end of the world to be in 1925. He later died leaving expensive homes for Abraham, Isaac and Jacob! While millions of people in the US were suffering in the 1930s Economic Depression, Rutherford lived an extravagant lifestyle in Europe and owning two Cadillac's. (This lifestyle is like that of the leaders of Hamas who don't live in Palestine but live lavish lives elsewhere). It is very sad to see our local people continuing down this false path and millions of un-informed people world-wide being led astray by the related false teaching like the Prosperity Gospel.

Several members of this group refused to be vaccinated during Independent Samoa's two-day mass vaccination campaign in September 2021, according to the Health Minister. From this, I now know that members of this group are much smarter than those who have been vaccinated including President Biden, the US CDC doctors, the LDS Church Leaders; Catholic Church Leaders; LBJ/DOH, Samoa medical personnel and thousands of doctors, scientists and researchers all over the world. Members of this group also refused to do blood transfusion[10], which makes it dangerous to be a member of this un-patriotic group. Later (Samoa Observer 2-10-21) a formal statement from an **alleged** spokesperson Sio Taua stated that the church supports the government's vaccination initiative

---

[10] Per. Comm Faletua Malaea S Brown 1989. Malaloa

and that most of their members, including himself, have been fully vaccinated. According to Taua the decision is a personal decision and a health choice. Talking about confusion within a church! but people can go to the www.jw.org to obtain more information. Samoa has a national Rugby Team called the *Manu Samoa* and some members of this church really enjoy watching this national team play at the international level, even though the church doesn't promote national pride. I know how some of them felt (national pride?) while clapping when the National Rugby team *Manu Samoa* in the 1991 reached the quarter finals after defeating Wales. I know this because I was watching the game with some of my friends who are members of this church.

According to the Daily Beast, 7/27/21, a new feature-length Vice TV documentary has set its sights on the Jehovah's Witness in regards to child sex abusers and the punishment of victims for seeking justice for their horrific ordeals. Aaron Kaufman's film *Crusaders*—released as part of Vice TV's "Vice Versa" nonfiction series—eviscerates Jehovah's Witness in which he was raised and providing a platform for members of the church to speak out including information about the elders who are committed to keeping these acts a secret. The *Crusaders*, builds upon Douglas Quenqua's 2019 *Atlantic article* about a secret database of thousands of Jehovah's Witness child offenders (created in March 1997) that has been assembled and concealed by the Watch Tower Bible and Tract Society, the not-for profit organization that governs the church. Whistleblower complaints resulted in questionaries being send out to 10,000 nationwide congregations asking for suspected church members being pedophile predators.

When Russia invaded Ukraine in 2022, many Ukraine men residing overseas returned home to fight for their country. I don't think these included members of this group. Samoans were appreciative when President Biden declared in early 2022, that no American troops will be fighting that war especially since American

Samoa has the highest number of soldiers in the US armed forces, **per capita**, than any US state or territory.

This Faith is well known for handing out tracts or literature and proselytizing door-to-door for many years. Its adherents have been required for the past century to make regular reports on how many hours they put into such ministry. Former adherents often tell of pressure to meet these quotas and guilt when they didn't. I know this to be true because I had a friend who used to do this and would often leave in the middle of a gathering with our other friends saying he has to meet his quota. This practice reportedly ended in late 2023 for rank-and-file adherents (Associated Press).

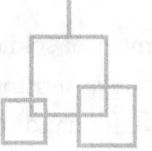

Assembly of God (AOG)

(Fa'apotopotoga a le Atua i Samoa)

In 1928 Rev, Herman Winkleman arrived in American Samoa and later Rev. Maurice H. Luce and Rev. Roscoe, to start this church in these islands. In these islands, the church was formally established around 1949 in Pago Pago. The church grew and spread over to independent Samoa. In 1976 the first graduation of its Assembly Christian Training School (ACTS) Lepuapua, Leone held its first graduation. A more comprehensive history of this church in American Samoa is well documented in my first book: "*Tala Fa'asolopito o le Ekalesia Fa'apotopotoga a le Atua i Samoa (Assemblies of God, AOG) i Amerika Samoa*". WestBow Press 2012.

Seventeen members of the Capstone Assembly of God Church in Iliili successfully completed a "Basic Sewing" Workshop conducted by the American Samoa Community College (ASCC) – Agriculture, Community and Natural Resources (ACNR) Division's Family Consumer Science (FCS) Program on November 11, 2021. Participants learned about the fundamentals of sewing from the ASCC-ACNR FCS agents Diana Tarrant and Shalley Tailevai.

In March, 2022, *Faletua* Amy Wendt Tavai of the Alofa Tunoa Pentecostal Ministries passed away in American Samoa. She was my neighbor growing up in Apia and we attended the Malifa Primary School. She and her husband Bishop Dr. Rev. Elia Esera Tavai served their church for many years in Hawaii and American Samoa. She was buried at her home in Vaitogi after her funeral services led by the EFKAS pastor from Fagaalu.

AOG Pago Pago elected a new pastor in September 2023. Pastor Faraimo Pagaialii had served this church for about 20 years and was happy to continue serving his church, which I also attend while writing this book.

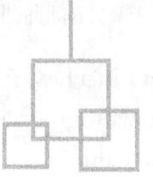

Bahai Faith

(Faatuatuaga Bahai)

E ven though this Faith doesn't believe in Jesus as the Messiah and Son of God[11], I've included it as it was a religion that the late King Malietoa Tanumafili II joined while he was the Head of State of the Independent State of Samoa. This Faith has more than 5 million followers over 300 countries and the lady who brought this religion to these islands, Lilian Wyss-Alai, arrived on Upolu on a boat in 1954. She was a former Presbyterian Sunday school teacher. The Faith started spreading in the US in the 1900s.This religion originated in Iran and has three fundamental principles, which establishes the unity of the human race or the oneness of humanity:

1. Unity of God
2. Unity of Religion
3. Unity of Humanity

These three principles' coherence concept may have resulted in the absence of its followers committing heinous crimes like flying planes into skyscrapers; evil acts which Professor John C. Lennox described as "blind faith". Funds to operate this religion are allegedly from its members and there are no collections during its services.

11

A service to mark the 60th Anniversary of Samoa's Independence was held at the Bahai Temple in *Tiapapata* in 2022. The event recognized the important role played by the United Nations (UN) in Samoa's March to Freedom. According to information provided for the event, "Samoa petitioned the UN for self-government in 1947. Six UN missions followed -1947, 1950, 1953, 1956, 1959 and 1961 until the UN General Assembly terminated the Trusteeship Agreement recognizing Samoa in December 1962 as an Independent state. It moved from a heavily Indebted Poor country (less that USD 500 per capita) to a Least Developed Country in 1971. In 2014, Samoa was classified as a Mid-Income Country and in 1976, Samoa joined the UN and embraced its values and principles.

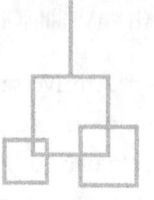

Anglican Church

(Lotu Anelikana)

T he All-Saints Anglican Church School at Malifa hosted a group of educators who came to strengthen partnerships with the school and also enhance networks in Samoa. This was according to the CEO of the Australian-based Anglican Schools Commission. The church had been sending young teenage volunteers from Australia for two-week periods to learn the Samoan language and culture as part of its cross-cultural experience. The young volunteers were also able to enjoy the natural beauty of the islands, experience the beach *fales,* hospitality and visit the Cultural Center at the STA premises, the Ocean Trenches and the Museum of Robert Louis Stevenson.

The church held a very successful bazaar during this period and collected around $60 thousand Tala. The funds were for the maintenance of the Leififi Church and the Vicars residence inside the church's compound. This was an annual event and featured children's' games, various food items and raffle prizes. His Excellency, the British High Commissioner was a winner of one of these raffles, winning a real live cow and the Director of SPREP won an oinking pig!.

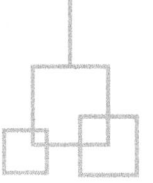

Church of the Nazarene

(Ekalesia Nasareta i Samoa)

In 2022, the government of Samoa gifted the Church of Nazarene with $20,000 Tala to mark the 60th. Anniversary of the church's establishment in Samoa. This church arrived in Samoa in 1962 and started setting up churches at Lotopa, Falelima, Ululoloa, Falelauniu and Vaitele. The celebration was held at the Church's headquarters at Ululoloa and was attended by many local government leaders. Overseas church dignitaries from America were also present. The theme of the celebration was: "to sail with faith". The church was involved in Christian programs such as Youth for Christ (Coordinator Jerrie Garsee) and Showers of Blessings. The Acting Prime Minister pointed out that the celebration coincided with the celebration of Samoa's 60th. Independence Anniversary.

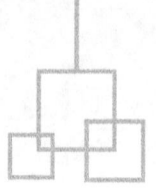

Church and Politics

The year 2021 will be in the history books of Samoa as the year when democracy was tested to its limits. Democracy and peace are always essential to the progress of Christianity. The HRRP (Human Rights Protection Party) party that has been in power for more than 30 years refused to give up power after the general elections causing a three- month long political crisis. (This is very similar to the situation in the US when Joe Biden legally won the Presidential Election and GOP supporters refused to concede). HRPP members refused to acknowledged the democratic process and refused to abide by the law of the land. HRPP was wrong as evidence by their formal apologies provided later, the consequent charges against some of their members and the resignation of some of the Party members. The new FAST Party, led by Samoa's first woman Prime Minister finally unseated a party led by a male Prime Minister for many years. FAST took office around July 2021. Politics often destabilizes countries, fortunately, the chaos that followed that election was finally resolved when the new Prime Minister took office and the country gradually got back on track. The change in government also brought out buried personal ethics and integrity issues when several HRPP members started thinking for themselves (not forced by the Party Leader[12]) and voted with FAST in regards to a new regulation taxing pastors. This was also evidenced when

---

12

some HRPP members resigned from this party in late 2022. This phenomenon is real in international politics. In September 2023, US President Biden said on 60 Minutes that there are a few Republicans who personally told him that they "agree with his policies but they couldn't openly vote for those policies since they will definitely lose their constituency elections".

Pondering and personally examining issues and not going with the flow can also be evidenced in many churches. A congregational church in Upolu started its bingo games and some AOG denominations on Tutuila ran weekend Car Wash to raise funds even though these were disallowed by the Board and Elders of the respective churches. Overseas, a Catholic priest, Frank Pavone, was removed from the Priesthood because of his outspoken work for the pro-life movement.

A US supreme court justice once noted in 2021 that "you have a choice between putting country over party" (Southern Bend Tribune. 4/4/22). The greatest blunder in party politics is when politicians vote along party lines instead of considering their constituents desires and their own personal unbiased and professional convictions. Samoa hasn't seen so many pastors in court since 1962, during this period, when several CCCS pastors went to court in regards to the new HRPP-introduced laws taxing pastors. A few pastors were in the news supporting the new HRPP tax laws but according to my interviews for this book, these were coerced and didn't reflect the opinions of the majority of parishioners and pastors of those respective denominations. Did the majority of voters thought taxing the pastors was a good and appropriate move?. or did it come about due to the iron fist of the HRRP leader?. It was certainly NOT the will of the people (comments by members who later left HRPP). This was proved when the Tax Amendment Bill 2021 repealed taxing the pastors and interesting though, some of the HRPP members voted against the former law. According to related statistics, 23 religious leaders paid taxes out of the 600 plus pastors in Samoa. The church has a lot to do with politics in these islands since all Parliament members (in Samoa) and House/Senate members (in American Samoa) are parishioners of some church denomination.

In the 1950s there was a major problem with the drinking water in France and thousands of people were affected. The problem was finally solved when a single major connection on one of the main pipes was **removed** and **replaced.** The **source** of the problem was identified, appropriate actions were implemented and the source of the problem was removed. Many problems can be easily solved if the source is identified and removed. Removal and replacing occurred in Samoa politics during this period and the churches were heavily involved in this event.

In Biblical times, the Sanhedrin's normally chooses the High Priests but the influence of the Romans sometimes seeps into this process where they advocate on behalf of a candidate that they can easily manipulate (R. Ken Suito). This is similar to the situation when one Cohn gave biased advice to a former US President (FRONTLINE. US) Churches in the Samoan islands have generally stayed away from politics up to the early 1960s. Governments have however requested the churches to participate and solemnize many national events in the past decades. Formal openings of its legislative assemblies have always started and ended with a Christian hymn and a prayer. The National Anthem of Samoa, composed, by Sauni Liga Kuresa, adopted in 1962, a few yards from where I grew up in Apia, contains the words *"Atua* - God" and *"Iesu* - Jesus". Maybe this is why the inhabitants of these islands are generally perceived as Christians.

American Samoa, a territory of the US, has found it interesting that many misguided televangelists in the US had publicly supported certain politicians. (Newsweek 11/12/20). I hope these US evangelicals would come down to *Falealili* so that the chiefs can teach them Politics/Religion 901 which teaches the stupidity of supporting a politician who advocates insurrection and defamation and that any individual or team vying in a sport or for public office are **all** God's children. Therefore, the churches should never ever show preference by supporting any contestant in any type of race, but may pray for their safety and improved comradeship.

Major adversities have resulted from individuals of certain denominations involving themselves in politics. For example, in 2022 a Sydney-based Samoa radio host was charged with one count of defamation and he later apologized to the Methodist Church for tarnishing the name of the church. Apparently, this person was attending the Church's Annual Conference in Samoa in 2022 but had allegedly defamed the FAST party with unfounded allegations and apparently failed to enroll in the *Falealili* Political Course mentioned above. Apparently, the good broadcaster and member of the clergy was later acquitted by Samoa's High court but not by the court of public opinion. (During this period an Ohio pastor was arrested for his alleged role in the US January 6, insurrection).

Related atrocities had happened in the past in foreign countries when government partners with churches. For example, in Canada where a social program that was aimed at "Killing the Indian and Saving the Human" was sponsored by the government and the Catholic Church. The 2021 investigations by the *60 Minutes* television program suggested that hundreds of children were physically, and sexually abused and died in the schools under this program.

In 2021, several parishioners were called to testify in various court cases related to national elections in Samoa. The issue of donations to the churches was also examined by the court during these cases. The practice of *monotaga* and the related *matafale* were queried as these concepts and practices were brought in by opposing parties to justify respective candidates' qualifications to run for parliament. Related testimonies revealed diversities and variations in the communities' perspectives and interpretations regarding these practices. For example, some churches believe that these traditions are compulsory. Several CCCS Pastors were also asked to testify.

In relation to the recording of Samoan traditions, there is a bias in interpreting some early information after the arrival of the

London Missionary Society missionaries in 1830, because most of these were recorded by LMS missionaries and German authors. Later interpretations of Samoan sayings, without the Christian bias, reflected some more secular clarifications. For example, the saying regarding *"o lo'o ao pea le masina…"* was explained to me by some old chiefs as having to do with the male genitals. I thought this was ridiculous and believed that they were joking. Twenty years later, the chiefs explanation was confirmed by two National University of Samoa (NUS) Instructors. This is why it's important for researchers to refer to the books 'Samoa Islands" Kramer series for additional information about these islands.

In 2020, the President of the National Council of Churches warned parishioners not to get involved in politics. Earlier, the General Secretary of the CCCS contradicted this opinion when he said that the church should be involved in politics. According to the President "the role of the clergy is to pray and fast for the spirit of the Lord to guide the voters in making their own rightful choices on who the political leadership they believe and trust". Now I completely lost my trust in this General Secretary. There were several related conflicts all around the islands during this period and members of the clergy also provided capricious perspectives on this issue, adding more confusion to unsaved and indecisive souls like me and my professional musician friends.

During this period, Radio Canada's French service reported a "flame purification ceremony" conducted by the Conseil scolaire catholique Providence, which includes 30 schools in Ontario. In 2019, 30 books were burned and their ashes used to fertilize a newly planted tree. A further 4,700 volumes have apparently been removed from the schools' shelves and designated for destruction or recycling. The organizers of the burning don't see themselves as heirs to Nazis. To the contrary, the incinerated books were chosen for their ostensibly degrading accounts of Indigenous peoples. According to a statement published in *The National Post*, the Conseil designed the ceremony with the participation of "many Aboriginal knowledge keepers and elders." The event predates the summer's rash of church burnings, which followed the revelation of hundreds of unmarked graves at state-funded, church-administered residential schools that operated from the late 19th century until the 1990s. But it is clearly "part of efforts to come to terms with darker episodes in Canada's past that parallel iconoclastic antiracism movements in the U.S".

Churches should not meddle with a country's politics because that's not their calling and churches generally don't understand the intricacies of political issues where the devil hides within the details and hidden agendas. Stay in your lanes pastors. This is not to say that the church shouldn't be involved in community work and government programs for the good of the public. Sometimes, participating in local politics reflects the pastors' immature knowledge of related issues. For example, in September 2021 when the HRPP members tried to go into the first sitting of Parliament and the police stopped them, a pastor "knelt on concrete outside of parliament pleading for the HRPP to be allowed to meet with the speaker" (*Samoa Observer*). Pastors should only kneel before the Almighty God and no one else. Isaiah 45: 23 doesn't mention or allow kneeling before a parliament, police or government leaders. If this was a pastor of the village of *Salani, Falealili*, I would suggest to the congregation and village chiefs to fire this pastor because this act has demeaned the high standard of the clergy (specifically the concept of the Christian God) within the village. A few members of National Council of Churches (NCC) later arrived to help resolve the conflict but members of the CCCS – the country's biggest denomination - were not present, reflecting this denomination's disgust with said political party. HRPP had taken pastors of this denomination to Court on tax related issues and this was truly the **start of HRPP's downfall**. The downfall of some leaders can sometimes be attributed to their failure to surround themselves with people much smarter than themselves. Well-known US millionaire Warren Buffet gave someone this advice: "In order for you to succeed in life, you need to surround yourself with people much better than you". A few years ago, Utu Abraham Malae, the only person in the Samoan Islands (**and most probably the only person in the whole world**) who was appointed Director to four American Samoa government departments at the same time (ASPA, TEO, Port Administration, LBJ), personally gave me this advice: he said to me: "Fui, my secret to success is surrounding myself with people much smarter than me and treating everyone well". Again, the local governments shouldn't meddle with church issues and vice versa. Does this suggestion apply to overseas countries?. In February 2022, the Pope made a surprise visit to the Russian embassy to express his concern regarding Russia's invasion of Ukraine that happened a few days earlier.

Related to the idea of revering members of the Samoan clergy was a barbaric event that occurred at *Lefagaoalii*, Savaii in September 2023, where an elderly man was beaten and tied to a stick by the untitled men of the village on orders from the Village Council, because he allegedly swore at a Catholic Deacon. Several people were later charged regarding this incident. Parishioners of all denomination should know that the Law is above church in these islands. In one instance, I think the clergy in the village of *Faleatiu* mistook the imminence and high regard for the church when a school building burnt down and the village refused to utilized the buildings that belonged to the church for the school. I think Jesus would allow the children to use the available church houses as long as they don't teach atheism.

According to a US Calvin University Professor of History, "When politics meet faith and Religion, people leave their faith behind". During this period in the US., Republicans threatened to impeach a Wisconsin Justice before she can rule on impending cases. Politics sticking its head into the legal system was also not a good sign for Pastor Stephen C. Lee who was indicted for racketeering in a case related to a US insurrection during this period.

In the US, Some of the Pentecostal pastors and evangelical advisors like Mike Evans, and Robert Jeffress who advocated for a former US President had later turned-on the former President during this period. Televangelists Paula White and Copeland who prophesized that a certain president will win the Presidential election should also apologize to the public and leave the Gospel arena for good because their false prophesies have proven they're fraudsters. I predict that these two will never repent nor apologize and if a former president loses again, it will be these preacher' fault. Charismatic preacher Kat Kerr said that God spoke to him and said that "He (GOD) will be in trouble if he didn't provide a fun place for kids". Can you imagine God being in trouble?. According to some Christian podcasts, some Evangelists had allegedly did forbidden acts with other women in the recent past. These preachers, as soon as they appeared on television after allegedly committing illegitimate sexual acts, said on television that the Bible teaches to never touch those anointed by God. After a meeting with the chiefs of my village, we unanimously agreed that these preachers and many others like them, were never anointed by God in the first place and are not welcomed to our village.

In 1967 the Shah of Iran held the most expensive party in the world where he crowned himself "King of Kings" because he believed that God spoke to him to do that. This was the beginning of his downfall and his gradual death. A street beggar named "Staka" when I told him about this event said that it was him that spoke to the Shah and not God.

Maybe pastors shouldn't meddle with politics as it may consequently reflect their shallow and incompetent resolves. From these events, I have concluded that there are no true Christian prophets in the world today, but there was one in Samoa during this decade. Her name is Toa and her prophecies were validated by several Biblical scholars and the Pope.(Read my previous book for related information). I've also watched a few videos related to this event, showing blood on Toas' hands, feet and head, and did some research and concluded that these events regarding Toa are **genuine and true**. A genuine Samoan prophet.

Many people in the US believe that the Climate Change issue is politically-driven according to a Pew Research Center survey and politicians try to attract voters who believe in the parishioner's God-given duty to be good stewards of Mother Earth. Some people in my village believe that God has a "hands-off" style of management and had a preference to "delegation" where He created Earth then delegate humans to care for it, a fact mentioned in Genesis. Genesis 1:25 also stated that "…..and God saw that it was **good**". Therefore, whatever happens later (after Genesis 1: 31), is of course "not good". Adam and Eve were however, also initially "good" creations but they were also given freewill, which resulted in the "not so good world" thereafter, up to this day.

A year after US troops left Afghanistan in 2022, more than 40% of the population were living on less than one meal a day and 97% were expected to be living under the poverty line by the end of the year (NBC New World.8/15/22. Hyder Abbasi). The religious beliefs of these people, at least the leaders of the main religious groups fighting for dominance in this region, unquestionably impacted the wellbeing of the public in a very negative way. Prejudices, **religion** and bigotry can seriously impact people's wellbeing and people should just drop all of these ideas (including religion in general – but not true Christianity) and put the wellbeing of their families first. I know I would. The Congregational churches practice of *Fa'amati* and *Faigame* in some Samoan villages have been a serious financial burden to many families in these islands. Apparently, these church donations don't guarantee one's admission to everlasting life. God wants the return of one's spirit, not the return of one's money - which is his money in the first place. According to Dr. Jay Smith, Islam has resulted in the poor conditions in many countries in the Middle East (refer to his presentations on the internet).

In these islands, misinterpretation of the Bible in addition to denominational doctrines and practices that are not in the Bible (I thoroughly researched this issue and these types of church contributions are not in there) contribute significantly to poverty in the villages. This is according to several interviews I conducted for this book, experience and also from my research. I also personally observed a related trend when I walked the streets of Apia in early 2023 and witnessed the marked increase in the number of beggars and the number of children vendors on the streets. This was despite novel government and NGO programs for these indigent people. Interesting though, several respondents to my interviews were reluctant to blame their churches for the significant monetary contributions "required" of them. The cultural paradigm of respect, obedience to the chiefs and the shame if monetary donations are not enough, unfortunately comes into play. A couple of people who live near Apia told this author that their family has joined the LDS church because that church doesn't require hefty traditional contributions during funerals. I can attest to this real trend since this happened to a couple of families I know and also to my niece[13] who lived in Apia.

Sometimes, churches become involve in matters that relates to politics and culture. They shouldn't. Sometimes politicians cross over to deal with local legal matters. They shouldn't. During this period, the US GOP kept bringing up politics even though, the indictment of a former President was a **purely legal matter.** One of the disgusting aftermaths of this intrusion into unsolicited areas is that followers get 'radicalized", become insurrectionists, and later jailed. In Scotland a related matter[14] resulted in some homeowners being bullied into selling their homes so that a Golf Course can be extended. Scottish officials later opened their eyes and stated that they couldn't rule out the use of Eminent Domain, should the owner seek future expansion of his golf course. Sometimes, religion hinders social wellbeing of communities, like in Sudan where Sharia law has angered thousands of people, mainly women. Related incidents have occurred in these islands.

13

[14] Journeyman Pictures. Anthony Baxter.

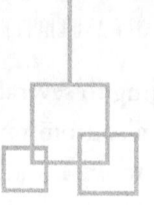

Church and the Law

T he Torah, the first five books of the Hebrew scriptures (Pentateuch) has an expansive and comprehensive list of "laws" that pertains to nearly every aspect of everyday life. It has more than 600 of these, directed at specific activities. For example, there are "laws" regarding the removal of eggs from a bird's nest and of course the conventional commandment to rest on the Sabbath. Christians should also be reminded about Mathew 5:17 "Do not think that I have come to abolish the Law or the Prophets, I have come not to abolish but to fulfil them". No human being has (and will) ever be able to practice and faithfully observe the law of Moses, with all its hundreds of precepts. Only Jesus can, but then he worked during the Sabbath, because He is above the law and the Biblical laws are for humans. In these islands during this period, prisons were crammed and the churches need to play their part to mitigate related problems.

The US's Innocence Project has done a lot of good for the innocent people (mostly Blacks) in the past several years. This project has resulted in more than 300 ex-convicts released from prisons. Hundreds more are possible candidates and hoping for justice, which would never be sufficient until all the corrupt prosecutors, attorneys and policemen affiliated with these cases are send to jail for a mandatory five years, awaiting the Last Judgment mentioned in the Bible. The two Samoa Governments do not have such a project but there should be one with the goal to review past Land and Titles court cases, where injustice was overwhelming and many families lost *matai* titles and hundreds of acres of family/communal lands. There are a several of those including a case related to my family lands at *Malie*, which we eventually won (thanks to Maualaivao Faualo Pepe Seiuli) after many years of court appearances. The names of *Molioo* (American Samoa Court employee) and *Gurr* (Samoa and later American Samoa Court employee and was expelled from Apia courts) seem to come up often during my research for this book, in relation to unethical problems with communal land registration. This type of injustice can be corrected and has been done. In California, ninety-eight years after officials seized prime oceanfront land from a Black family, a Los Angeles Commission voted in 2020 to return the property to the original Black owners. During the Black History month of 2023, it was reported that scores of families in the US are fighting to get their lands taken from them. Samoans should vigorously fight to do the same but it seems that several related land cases have not resulted in positive outcomes for the traditional land owners. Where are the native Samoan attorneys?. If the Attorneys in the US can bring back lands to their original owners, why can't they do the same for their local community?. Well, they might argue that they need related laws to do this. The public can easily talk to their *Fono* members and I'm sure they will help as some of their family lands are affected. This is the type of issues that our local students were sent to Law schools for. Villages, like *Aoloau*, should vote in warrior legislative representatives that are brave to introduce laws to cater to this problem. All the churches are built on donated communal or family lands and that is one of the connections of this issue to this book. Reflecting on this issue, the Samoan legal system is more enhanced than those of Louisiana and Ohio in that these places have judges (not Associate Judges) that have no legal training.

There is a related stupid law that exist in Texas, called the "promiscuity law", where modern forensic science later proved that the perpetrators who were exonerated using this defense should have been charged in the first place.

A historical moment in the history of the US legal system was attained in August 24, 2023 when a mugshot of a former President was shown on national television. Three hours later, I watched a video presentation about several US judges being "jailed" in various US States. It's a gratifying feeling to know that in the US Legal System, no President nor Judge is above scrutiny.

In another related issue, a US judge ruled in 2022 that a Tulsa Race Massacre reparation lawsuit may proceed. The initial event happened in the Black Wall Street where several prosperous black residents were killed and their homes demolished by white supremacists on May 31, 1921.

Circa November 2020, Samoa's Court ordered the Police to remove a pulpit from the Amazing Love Christian Church at Samatau, following the death of a three-year-old toddler early in the year, when the pulpit fell on the child. The judge said that it was inappropriate to continue to use it and the church should have looked into this to avoid similar episodes. Churches are not always safe places, not only in Samoa but also in overseas countries where several people were killed inside churches and mosques. In Vailoa Samoa, a lady was punched in the head by her boyfriend in church during this period. In November 2023, a Mr. Taotofi was arrested at Vaitogi, American Samoa, after he tried to fight with his pastor inside the church.

At least 41 people were killed after a fire at the Egyptian Coptic Christian Church, in the city of Giza, most of them were children[15]. Many others were killed in US churches while they participated in Bible studies and church services in the last decade. A Catholic bishop was fatally shot in Los Angeles in February 2023. All these events happen while many others were enjoying drinks at the bars.

---

[15] Reuters.2022. Hassan. A.M., Sheasha. Sayed.

Where would you prefer to be? ask your pastor or look for me at both places!.

In March 2022, President Biden announced his nomination of a women of color (Ketanji Brown Jackson) to the US Supreme Court. She will join a court made up of six Catholics, a Jew an Episcopalian who was raised Catholic. This is very important as the judges' votes tend to be biased towards their personal religious beliefs. Later in April 2022, Jackson was confirmed by the US Senate. This is very important because American Samoa is still (and will always be[16]) a US territory, and later decisions, like Roe v. Wade was; was later decided by the Catholic majority in the US Supreme Court. Some believe that the US Constitution should remain a philosophical document but should have several interpretations and a few legal implications. Justices should be philosophers first, then attorneys to keep the "intent" in the Constitutions' preambles alive and applicable. Yes, the church can have a significant influence on government, politics and justice, but in June 2022, it seems the US Supreme Court's Decision to overturn its 1973 verdict is now, significantly influenced by religion. I think that the US supreme courts should employ only atheists as Supreme Court Justices and maybe one Buddhist so as to prevent religious bias in their decision. Since it's a medical fact that human cognitive abilities decline with age, an age limit of 70 is appropriate for the justices. Maybe the Justices of the US Supreme Court should be voted in by Congress and not by the sitting President. A ban on lobbyists inside the halls of Congress will also help. I plan to ask my friend F.A. Utu (American Samoa Attorney General) for his perspective later.

During this period, Samoa's parliament passed three laws that would fundamentally altered the country's constitution and judicial system. The proposals were widely condemned by lawyers and judges and resulted in Fiame Naomi Mataafa leaving the HRPP. The new laws elevate the Land and Titles Court into a stand-alone judiciary

16

equal in standing to the Supreme Court, which removes the high court as an avenue of appeal. This was a very dangerous and stupid route to follow according to those interviewed for this book and the Law Society. For any issue, people should heed the advice of experts, and in this case, the experts are the local judges and attorneys. These new laws were seen by the judicial community as "ill-conceived and poorly drafted"[17]. Many families, future generations and of course parishioners of all denominations will be negatively impacted by these new rules.

Israel also introduced similar laws during this period which resulted in mass protests, and was struck down by its Supreme Court. The will of the people must be heard, not the determination of sitting governments.

The Independent Seventh Day Adventist Church in Auckland (under the leadership of Pastor Papu in 2021) is not the only church that was investigated by authorities during this period. The Mormon Church settled (abcNews K.Aaron. 2/21/23) accusations it covered up $32 billion investment portfolios in 2023.

The son of a Methodist Church minister appeared before the Supreme Court of Samoa in early 2023, allegedly charged with rape, and there was a noticeable presence of members of this church in court. The accused was referred to Legal Aid as he was unable to afford an attorney.

Around April 2023, the Legal Communities of Samoa and New Zealand met in Apia to discuss the traditional *ifoga* as it applies to the law. Various related views on this issue were expressed during this period. According to some people and NGO's, the traditional practice of *ifoga* shouldn't be acknowledged in cases involving child and sexual abuse. I totally agree with this perspective. All perpetrators of these crimes should absolutely be jailed and the *ifoga* shouldn't affect their sentences. *Ifoga* is a cultural concept and shouldn't be acknowledged in a court of law that is based on western legal

---

[17] Lesatele Rapi Vaai. President of Land and Titles Court, 2022.

concepts. Additionally, the acts of these perpetrators are not only against Christians beliefs, but also against US inmate persuasions where scores of pedophiles are beaten to death by inmates.

The Samoa Victim Support Group holistic preventive and responsive child protection program, in partnership with UNICEF held its community program in *Toamua* in May 2022. The program is to try to prevent child abuse through ensuring that "communities have better understanding of violence against children and are able to prevent and respond to the same". Other villages targeted by the program include *Samalaeulu, Siumu, Solosolo, Luatuanuu, Fasitootai and Fagalii.*

There is a relatively new practice that seem to help poor people fight the big companies in US Courts. The Litigation Funding. Investment Firms who run this practice would pay for the litigation on behalf of the "little guy" against big companies and have mostly won most of their **selected** cases, gaining millions of dollars since the last decade. Helping the deprived people is a biblical concept. Making money out of that process is up for debate.

In August 2023, a Samoa church minister accused of allegedly raping a 15-year-old girl secretly left the country while the matter was being investigated by local police. According to police, the pastor, however, has not yet being charged. The victim, who was placed at the Samoa Victim Support Campus was reportedly, removed from this safe niche by his uncle and the victim's family was banished from the village.

Watching many debates between Christians and atheists, I've come to appreciate one attorney's advice to his client who was accused of failing to restrain his dog from biting the plaintiff. The attorney told his client not to worry because he has a four-fold defense:

1. The dog is on leash all the time
2. The dog doesn't have teeth
3. The dog doesn't bite
4. **He doesn't have a dog!**

You the reader may interpret this illustration, if you can't, let me know and Ill refer you to my friend Fainuulelei-Utu. For people who don't know much about the law and would be very nervous when stopped by police, one lawyer gave this advice: whatever question the police asks, just give these two replies:

1. I don't recall
2. I need an attorney

Ponder this true story. About ten years ago, a *faletua* was stopped by a policeman at *Leloaloa*, American Samoa. When she asked why she was stopped, the policeman said that he thinks she (driver) is drunk. The *faletua* smelled alcohol from the policeman and she replied, I think its you who is drunk. The policeman, a very good friend of mine, let the *faletua* go and now he's preaching the Gospel of Jesus Christ.

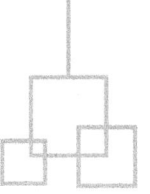

Church and the Samoan Culture

There are advantages in utilizing the Samoan culture to empower and elucidate church programs. For example, many Chiefs would encourage (sometimes force) members of his family to join and provide support for his church. Un-titled men often stand guard during church activities and evening curfews (*Sa*) for family prayers. Chiefs would always accompany a member of the family who is been called to lead a church, to perform traditional presentations. Law enforcement sometimes call on the Council of Chiefs to intervene in major village conflicts, and pastors would often participate in these conciliations.

Samoa has its own several traditional gods like *Tagaloaalagi*, *Lesa* (god of agriculture); traditional "deities", demons (e.g., *Moso* and *Nifoloa*) and idols like *Taisumalie*, *Alaiava*; demi-gods like *Nafanua* and *Pili;* ghosts like *Saumaiafe* and *Telesā* and other traditional entities before Christianity was introduced to our islands.(Cook Islands has a simila god Tangaroa and Hawaii has Kanaloa – god of the oceans and winds). A reminder to readers that pastors in these islands have proven in several occasions in the villages that the Spirit of the Christian God is much more powerful than these traditional entities. I've recently wondered why the Christian Faith has grown full-fledged in these islands (thousands of miles from Jerusalem) while thousands of people in Israel, (more than 90%) the original place of Christianity, don't even believe in Jesus Christ?.

Does Christianity has concepts similar to Samoan traditional beliefs and legends?. Yes. For example, some LDS members believe that heaven is made up of different tiers and Samoans traditionally believe *Tagaloaalagi* lives in the upper-most tier of heaven. Similarly, people of Mesopotamia and the Hittites also have their own their own god, Anu, the first known god in the cosmos who had a son name Kumarbi.

In the Philippines, mass shootings continued to be a problem but a cultural practice called *hiya* – a Tagalog word meaning shame or embarrassment, had deterred this problem significantly. When a Philippine is jailed for shooting another person, the entire clan will all feel humiliated and therefore many would-be mass-shooters, refrain from attempting to shoot other people to curtail embarrassing their families. Similar feelings have been expressed by some local rogue youngsters.

Several years ago, I participated in an American Samoa Community College (ASCC) awards ceremony where my granddaughter Finiana Tamasailau received a Science Award. The organizers of the event asked me to say grace before the diverse audience partake the food. I stood up and made the remark that since God is Samoan, I will say grace in Samoan. I had thought I would write a book about the Samoan culture and Christianity later-on with the title God is Samoan. Several months later a scientist friend of mine, Doug Fenner emailed me about that comment. Later, I found out that an anthropologist, Dr. Mathew Tomlinson had written a book titled "God is Samoan". I hold that this was originally my idea and emailed Tomlinson about my comment. Tomlinson later congratulated me when I published my second book. Tomlinson's book is *"God is Samoan: Dialogues Between Culture and Theology in the Pacific"* explored the "intersection of Pacific Islands culture and Christianity and how it shapes theology". This book is great reading for any Bible College student and a copy is always on my office desk.

For several years, I had an internal personal conviction that: speaking of the Christian God, humans don't know sh*t., but

I was hesitant in actually writing this thought into this book. I thought since this book is mainly about Christianity, it would be inappropriate to include this statement. Years later, I found out that a former Minister and author, D.B. Ramsey had the same view, so here it is, from me: Speaking of God, humans don't know sh*t. In relation to the origins of the universe and Christianity, readers can watch the video "The Poetry of Reality" where it was said that physics can explain time beyond time, but later on the physicist stated that "We will never explain all the laws of nature and there is no way out of that". When Professor Lawrence M. Krauss was asked about this subject he replied "we're happy to say we don't know". Richard P. Feynman wrote: "I don't feel frightened not knowing things, by being lost in a mysterious universe without any purpose, which is the way it really is as far as I can tell". To sum this all up, humans don't know sh't about God and the origin of the universe…. well maybe just a little. *Tulou lava Samoa.*

In September 2021 a National University of Samoa (NUS) Lecturer became the first Samoan to Publish a thesis **in Samoan** on Customary land and ownership and also became the first to be awarded a Doctorate in Samoan Culture. His research highlighted changes in customary lands from 1845 to 2020, specifically within the villages of *Amaile* and *Samusu Aleipata*. Most, if not all of the lands on which church buildings are built, including lands in these villages, are donated by church members and families of the respective villages.

The Catholic Church has been the prominent church that widely include some reflections of the Samoan traditions in its church services, during the past two decades. According to some Catholic parishioners, this is attributed mostly to prominent church leader Rev. Kasiano Leaupepe. Other denominations are also increasing their appreciation and inclusion of some of the traditions in their church services. In 2022, a EFKAS in Manua included the following Samoan traditional song; thanking God, towards the end on their *faamanatuga,* televised by KVZK TV.

Ua faafetai ua faafetai
Ua malie mata e vaai
Ua tasi lava oe
Ua tasi lava oe
I lo'u nei faamoeoe
(Translation by Daiana Anuilagi Peseti)
We thank you; we thank you
From what we've seen, we are very appreciative
You are the one, you are the only one
In my aspirations

During this period, a couple of Samoan Bible instructors hinted on blending into Christian teachings, certain parts of the Samoan Culture. I'm not sure of their perspectives but I suspect it may have something to do with some similarities or parallels between some Samoan Legends and a few Biblical stories. For example, the Samoan legend about *Sina*[18] (the legendary Samoan beauty) and the *tuna* (eel)- Kiribati has a similar legend naming its twisted island after their eel legend. The eel was supposedly the highest chief of Fiji who was transformed into an eel so he could swim to Samoa to seek *Sina* and take her as his pride. The *tuna*, in his last pronouncement, requested:

> *"Please cut off my head and bury it in front of your fale. A certain tree*[19] *(cococunt palm Cocos nucifera) will eventually grow, its fruits will show my face and whenever you drink the juice from this tree's shell it will be like kissing me, and you will then remember me for the rest of your life.*

If one removes the coconut husk, one could see the "face" of the Fijian. This is an **act of remembrance - communion?.** The parallel

---

18
19

concept is not new. To facilitate the understanding of Christianity, Saint Thomas used it to explain the scriptures to the thousands of people in India who had many traditional "gods" of their own.

The coconut tree is known by Samoans as the **Tree of Life** because of its abundant uses. Its leaves are used for weaving various baskets and rooftop covers for traditional houses; the leaves fronds (*tuaniu*) are used to make brooms; its shell is used as wood for fire and cups, its trunk is used to build houses; the husk is sometimes used to make the *tauaga*, to squeeze and strain out the white coconut milk from the kernel used in many traditional foods especially the *palusami*[20] (traditional "cake" made of taro leaves and coconut crem baked in the *umu*) and its juice is a delicious and a very healthy drink.

Another Samoan legend tells the story of a canoe that was built for *Tagaloaalagi* (God of Polynesia) daughters *Mataiteite* and *Matatalalo* by heavenly beings send to Earth by *Tagaloaalagi*. The village people were not allowed to see the canoe builders. One day, the village women, bringing in some food for the canoe builders, decided they will silently creep up to peek at these mysterious celestial builders. The builders were taken by surprise so they hid and later return to *Tagaloaalagi* in heaven because they didn't want the women to see their nakedness. Is this similar to the Genesis story of Adam and Eve hiding when they discovered they were naked?. Similar concepts but not parallel.

During traditional funerals; a *taupou* (family or village princess) usually sits beside the decedents body with a traditional fan –*ili*- to whisk away flies. This is called *toto'o*. According to a chief from *Falealili*, the Bible, mentioned that when the women went to visit Jesus's tomb, two angels were performing the *toto'o*. Maybe the angels were Samoan *taupous?*. According to a recent poll, 74% of Americans believe in God while 69% believe in angels.

The story in the Bible about Jonah being swallowed by a big fish

---

20

often invite incredibility from atheists. There is a Samoan legend about how the name Samoa came about. The story is daintily described in an old Samoan song about a couple, *Tufu and Lele*. The song has more verses than any Samoan traditional song, and has a verse that tells of how *Tufu* rode on the back of a shark. The relevant line in one of the versus goes like this:

Tali mai le malie ose mai e ti'ti'e I lo'u tua

Translated: The shark replied: "come and ride on my back". (Fortunately, the shark didn't swallow Tufu.)

There are also traditional songs that contain historic information. For example, a traditional song with the following lines:

Lau Afioga e le Kaisa
Se'i ola tama na fa'afolau

The word *Kaisa*, transliteration of the word Kaiser, refers to Kaiser Wilhelm who was in charge of Germany's drive to secure territories around the world in the late 1800's. The Germans landed in Samoa and made it a German Protectorate from 1900 to 1920.

A Samoan saying that goes: *"Ia so'oso'o le fau ma le fau"*. The concept is very similar to the idea in Mathew 9:17 "Now, they pour new wine into new wineskins, and both are preserved". The Samoan concept is to use the same material that the damaged item is made with to make a new bond or connection. Samoan orators also use to say during their speeches that *"ua paia ai nei maota ma eleele sa o le alalafaga* (the dwellings and all the lands of the village are now sacred – because God is present). Samoans can relate this to Exodus 3:5 "*....for the place whereon thou standest is holy ground*".

The Samoans are not the only Pacific people embarking on this relatively new idea of interlacing Christianity with culture. In 1964, coconut juice was used and served for a Holy Communion service in New Caledonia where photographer John Taylor took the photo.

Bread was served, during Biblical times, in about 90 percent of family meals in the Holy Land. This maybe the reason why bread is mentioned in the Lord's Prayer: "and give us our daily bread …". An example of the influence of culture on religion.

Traditional community behaviors also affect religion. In Biblical times many men and women would come into contact with their future partners at the wells when fetching water. I've known several Samoan couples who met at church (spiritual well) and later got married.

One central model in Christianity is the concept of sacrifice. In an often-told Samoan legend, then King of Samoa, Malietoa, practiced cannibalism for several years. One day, warriors named *Pa'ialala* and *Pai'atea* came to offer one of them for the king's meal. On their way, the two continued to argue profusely who will be the sacrifice. The argument got really noisy and was heard by *Malietoa's* son *Poleuligaga* who was nearby. *Poleuligaga* asked the men what their argument was about and was told that each one of then wanted to be the sacrifice so that the other one can live. *Poleuligaga* then offered to have himself as the human sacrifice for the king. The brothers wrapped *Poleuligaga's* body in coconut leaves and presented it to king *Malietoa*. When the sacrificial item was opened, *Malietoa* was astounded to find that it was his son that was offered for his daily meal. *Malietoa* immediately ceased his practice of cannibalism and these islands later accepted Christianity. The brothers on their way back home were happy and relieved. They jumped into a small pool of water, the one that can still be seen in front of the Malua Theological College (MTU) church building (see photo below) called *Tofuol*; - translated "diving when receiving life"; a living symbol of uncivilized sacrifice that has gyrated to western theology. In the past, the Fiji Islands were also known as the Cannibal Islands. Readers should note that there is a phenomenon called "galactical cannibalism" that occurs in outer space.

Other religions in overseas countries have parallels to Christianity and their own traditions. Tao religion on the other hand has parallels not with traditions but with quantum mechanics!.

In these islands, the churches are sometimes asked to contribute to discussions on social issues. During this period, the *Palauli* 1 District Development Council sought input from their district clergy in addressing the prevalent social issues hindering the developments at the grassroots level. The Council invited 19 *Faifeau and Faletua* (church ministers and their wives) from all denominations in the district to an evening of *Talanoa* session in August 2023 at the *Salailua Metosisi* Hall to do just that. The issues discussed included domestic violence, youth crime, illegal drugs, human trafficking and assistance for people with disabilities.

There are some positive effects of Christianity in the Samoan communities, but these shouldn't be seen as exclusively an effect of Christian teachings but rather of basic morals and common sense instilled in most humans. For example, the push towards eliminating traditional practices such as:

i.   *Poula* -where men from another village would end up running away with women from the hosting village, after an extended night of traditional dancing and entertainment.

ii. *Ati ma le lau* -Banishing villagers from their homes after a decision by the Council of Chiefs

iii. Barbaric invasive determination by a *tulafale* if a bride is a virgin before a wedding.

iv. *Tini* - the discourteous and offensive language expressed by a large group from the groom's family against the bride's family in the early morning of the wedding day.

Although there are parallels between Christianity and the Samoan culture, these are fundamentally two different "animals". The contemporary idea of replacing the word *Ieova* in the Samoan Bible with *Tagaloalagi* maybe acceptable to some local theologians, and could have originated from the idea that the God in the old Testament is the Hebrew God while *Tagaloalagi* is the God Polynesia.

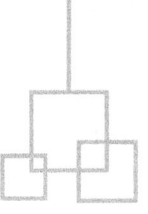

Human Rights

Traditionally, human rights have little importance in the Samoan community. People in villages seem to have modest rights as related issues are decided by the Chiefs. Evidences are scarcely mentioned – if there are any -, and witnesses are not often called to testify to the Council of Chiefs. Customary "sentences" are often extreme and the poor alleged abuser's family is expected to provide excessive food items, often disproportionate to the "crime". But that is the Samoan culture. Through the eyes of the Western cultures, this is unfair and probably uncivilized. Evidently, this is my culture and it is what it is and I'm proud of my culture. Many cultures of overseas countries have relatively more unjust and brutal rules. Readers should understand that Christianity and human rights are foreign concepts – not Samoan. The local community's inevitable progress towards the ever-changing socio-economic and political future might require adapting our traditions and perspectives. I believe related adaptations are critical to the survival of our *faasamoa*, and so we must ponder this issue and offer propositions on how our future generations might approach this and other related issues. The rule of law seems to be the right approach, and with the participation of churches and NGOs, these islands will be in a much better position with regards to this issue. Meanwhile in early 2023, Associated Press reported that according to the United Nations

Organization (UNO) Afghanistan is the world's most repressive country for women. (Associated Press. UNO. Faiez, Rahim. 3/2/23)

Maybe US troops should've stayed there for another twenty years to assist with women's rights with financial backing from the UNO?. Not a good idea since history has taught us that national changes have to come from within. Ponder the failure of Russian and US military in Afghanistan after many years fighting the local tribes.

In November 2021, Prime Minister Fiame Naomi Mataafa clarified that her new government will not change the laws only to satisfy some of the recommendations by the UN Human Rights Report. The rights emphasized in the UN report included euthanasia, same sex marriage and abortion and these issues were raised in the Universal Periodic Review of Samoa's State of Human Rights examined by the UN Human Rights Council (UNHRC). According to the Prime Minister, these fundamental rights are not only well-protected under Samoa's Constitution but any scrutiny should consider their relevancy to the Constitution and should be assessed in context with local traditions and customs. Some of these issues are discussed in this book.

The Samoa NGO Brown Girl Hopes, requested in April 2023 that a curriculum be introduced in schools to recognize human rights and the acceptance of the LBGTQ community. A member of this group had just attended the summit: Pacific Partnership to end Violence Against Women and Girls, held in Fiji.

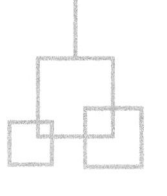

Translation, Transliteration, Interpretation and Linguistics

T he Samoan Bible should be the leading authority in Samoan translations because it is the first formal text that utilized the initial solemnize version of the Samoan language coordinated by the London Missionary Society (LMS) church. This idea was pronounced by the American Samoa Samoan Language Commission during this period, when it stated that "the Bible should be the classic standard" for standard language usage. As a licensed professional translator, I totally agree. l Church Sunday schools telling Bible stories in Samoan, the use of the famous *Pi Tautau,* and the continuous everyday use of the Samoan language by the two governments and most of the local inhabitants have all contributed to the solid unceasing eminence of the Samoan language. Related government-sponsored programs like the American Samoa's Annual Samoa Heritage Initiative include the Samoan language as one of its pillars. The American Samoa Department of Education's ASIA program has also promoted the use of the Samoan language in the past several years. In early 2023, it was announced that Samoan students at Stanford University, Ari Patu and Brandy Atuatasi, were successful in getting the Samoan language included in the curriculum and now being taught at the institution by mother-son duo Chris and Tali'ilagi Young. A 1956 documentary, *Samoa 1956,*

Dalk Broline CWD translated the word Samoa as "Sacred Center" and the word *taupou* as ceremonial host. As a translator, I find these translations interesting.

There are Softwares like Lingvanex, Google Translate and Quillbot, available on the internet that can translate English to Samoan and is generally satisfactory but shouldn't be used in a professional setting. I did a test and provided the following words in English and the Softwares provided the following Samoan translations:

English words provided to the software	**Software translation**	**The correct Samoan translation**
Divorce decree	*Tete'a Tulafono, tete'a*	*Tete'a fa'ale-tulafono*
Extended wisdom	*Fa'alautele le poto*	*Tofā mamao*
Come in your honor	*Susu mai i lou mamalu*	*Susū mai or afio mai lau susuga/afioga – note the use of the macron for the "u" which is essential.*

An interesting aside: *ichthys* means "fish" in Greek, but the letters themselves also form an acronym for "Jesus Christ, Son of God, Savior". In relation to "fish", some of Christ's disciples were fishermen and I did worked for the government's fisheries department (DMWR) for more than 20 years. It is also interesting that according to a relative of Bigham (former LDS President) there shouldn't be any translation problems with the Book of Mormon since it was originally in English, but apparently there are several (Aaron Shafowaloff; Sandra Tanner).Additionally, there are contradictions between sermons by Brigham Young and the text of the Book of Mormon.

I have **just a little bit** of knowledge and personal connection to this subject matter. My father, an LMS pastor, a Malua Theological College valedictorian, and a Papua New Guinea LMS missionary was the lead translator for the American Samoa Legislature (*Fono*) in the early 1970's and my brother and nephew were interpreters at the High Court of American Samoa at one time. I'm an American Samoa Licensed translator and had translated several materials for the US federal government in the past. My late uncle Liva Seiuli, was an editor for the Samoa Times newspaper for about 20 years.

Readers should be aware that there are variations of Bible versions call "textual Variants". These are words, verses, points where copies of the Biblical texts are not identical to one another. These variations occur as a result of copying newer ones from older versions and sometimes these can be minor or major differences, and sometimes these can be unintentional and sometimes deliberate. For example:

> "Christ bears (Greek: *pheron*) all things by the word of his power in most copies.

> "Christ manifests (Greek: *Phaneron*) all things by the word of his power (Hebrew 1:3) – according to Codex Vaticanus.[21]

The law is not just law but interpretations. The public relies on the Courts to interpret, but occasionally some bright attorney would come up with his/her own interpretation and consequently the actual intent of the law is then stretched to various dimensions. Legal opinions written by one judge, doesn't necessarily reflect the resolve of other judges. This is the same case with religion. The pharisees and other church leaders at that time have been providing their own interpretations of the Old Testament for many years and we now know that these were mostly inaccurate according to Jesus. This is exactly the same phenomena happening in Christianity in

---

[21] Trey The Explainer. 10 changes made to the Bible Part 1. 2020.

the recent past, where a variety of denominations provide various interpretations of the Bible; with some underscoring a few concepts and neglecting the essence of a topic, while some preach deceitful topics like the Prosperity Gospel then denounced it after several years of exploiting millions of unsuspecting followers. In the US, only a few televangelists that preached the Prosperity and Faith Healing gospel, later apologized in regards to their wrong interpretation of the Bible but then refused to return the millions of dollars they fraudulently received. In the past two decades, it's interesting to note that the indigents who collectively offered millions to televangelists and other sneaky pastors are partially to be blamed. The variety of denominations, doctrines and beliefs emanating from a historically single entity abound in overseas countries. In Iran, there is differences in convictions between Shia and Sunnis and as a result there seem to be a Shia Islam and a Muslim Islam. The normal conviction of Muslims that there is only one Quar'an is false according to Dr. Jay Smith who stated that there are about 26 different Quar'ans, and he also provided proof in 2016 at the Speakers Corner.

Here are a few interesting and related perspectives. A few Bible scholars suggested that Jesus's profession was "*tekton*". The most common translation is "carpenter" but these scholars suggest that the word "*tekton*" refers to craftsman or mason and Jesus was most likely a mason not a carpenter. If one says that his job is to "shoot people", it may mean that he/she is a photographer and not a murderer! Symbols seen in many churches can also be misinterpreted. The infamous swastika used by the Nazi Germans is actually a symbol of peace and prosperity to the Buddhists and Native Americans. The majority of early Christians outside of Judea and throughout the Roman Empire did not know the Old Testament in the original Hebrew but rather a Greek translation known as Septuagint, and Mathew quotes Isaiah 7:14 and uses the Septuagint term "parthenos" commonly understood as "virgin". The Hebrew Old Testament uses the word "almah" which can properly translated to ":young women". Caution should be taken when deciphering the Bible. Readers may also be interested in another perspective regarding Jesus. According to the "Lost Gospel of Thomas (discovered in upper Egypt in 1945 and likely written in the second century), Jesus is a spiritual leader rather than a divine figure.

Errors in translation and interpretations can have serious consequences and can lead to unbelievers questioning the stories in the Bible. Linguistics should also be examined to provide better explanations regarding Bible stories. For example, the words YAM SUF refers to the Red Sea area where the parting of the sea supposedly occurred (TUF = SUF in Egyptian). Archaeologists have for many years failed to find any evidence of this event in that area. About 40 years ago, some researchers claimed that the correct translation should have been the Reed Sea not the Red Sea. Performing archeological work on the El Balah lake, which has many reeds in its marsh waters, they finally found several scientific evidences of this event.

Words in manuscripts and books should also be properly defined, spelled and put in context to provide clarity to the readers. The word Nazi was coined by the Bavarian journalist Konrad Heiden, and meant "simple minded" and originally came from NASOS (National Socialist Party). These simple-minded group eventually killed millions of Jews (I'm sure some of those killed maybe distant relatives of Jesus!). A secular example includes the word Romania which should be Rumania! A related and interesting observation is that most of the stars we see at night have Arabic names because in the past, Bagdad was the intellectual center of the world. Unfortunately, during this period, it become the center of terrorism. One ox is one ox, so many ox(s) should be oxens?; 1 fox should be foxen? no, its foxes. The first human word spoken from the moon was "Huston".

Another related subject is the misuse of words in the Bible. Bible self and group study and the use of several resources available on the internet (including Hebrew and Greek translations) and praying for the help of the Holy Spirit, will solve realted problems. Interested readers should also read on concepts like typology which can have an impact on the interpretation of old Biblical inscriptions. According to Hebraist (expert in Hebrew) Steven Boyd Ph.D. of Hebrew Union College, even reading the original Hebrew version of then Bible, people can have various interpretations especially when people impose and add to concepts in the text. He also stated that there is no Hebrew word for 'universe". Relatively new detail analysis of the use of words in the Book of Mormon by Dr. Edwin Patterson concluded that Joseph Smith was not the author of this Book.

To get the "best" interpretation of Biblical scriptures, one should use the Hebrew-Greek versions and maybe learn a little Aramaic. But even this approach can present various interpretations but at least the user gets as close as possible to the original text. This is why grievous interpretations of Biblical scriptures by rookie so-called "pastors" in these islands is threatening the truth of God's Word. Most of these people didn't attend any local Bible school and utilize only the Samoan Bible, which is **very inadequate** since they're preaching the most important message of this age. A very annoying but relevant event started around this period where several religious podcasts by Samoans started to pop up on YouTube. One very unnerving local podcast called "*Talitonu I le Upu*" by a novice "child" touched on the Samoan Culture and traditional songs in relation to the scriptures. I'm pretty sure his parents never had time to discipline this youth, and the superficial inconsequential and frivolous statements he made reflected the absence of any religious training, relations with the Holy Spirit, academic achievements and his arrogant demeanor. I would suggest to him to talk to Dr. Puni of SDA regarding the Sabath and to the Instructors at Malua, Kanana Fou and Piula Bible Colleges on all other subjects before humiliating his family further. This type of podcast ruins the respected reputations of local theologians and churches. The denomination related to this podcast should immediately disown this perpetrator. What about freedom of religion?. Well, what is preached here is absolutely non-scriptural. The presenter also prophesized that God will burn down the village of *Falealupo*. This village *aumaga* should pay him a visit since I also have relatives in that village and I don't want them to get burned. The obligation to study the scripture in detail (in a Bible College or by personal comprehensive study) and understand Hebrew, Greek and Aramaic (most of Jesus followers speak Aramaic) wasn't fulfilled by this rookie. Readers can refer to I Timothy 3: 6 "Not a novice least being lifted up with pride, he fall into the condemnation of the devil". Additionally, according to one Samoan podcast critic, "the Holy Spirit didn't speak to this guy". This disturbing podcast can be likened to the "Deliverance Movement" with its publicized and sensationalized televised services that doesn't' actually deliver anyone from demons. Compared this to very **private** sessions where Catholic priests, *taulasea*, or Samoan Chiefs (like the late Prime Minister Mataafa) actually drove out demons from those possessed. The fact that hundreds of Samoan students are attending local and overseas Bible Colleges to learn about the scriptures, while this deranged boy suggested that this is not essential, is absolutely demeaning to these students, their families and the whole Samoan community. A local Methodist pastor commented in late 2023 that many locals are going on Social Media with gossip and unsubstantiated information defaming other people, and this is very disrespectful to the Samoan Community.

I suggest that every pastor should at least try considering utilizing the following three sources for every sermon so as to bring out the most truthful and best interpretation of the gospel, and also with the help of the Holy Spirit.

A. Septuagint (Greek translation of the Hebrew Bible; 3rd. and 2nd. Centuries BC).
B. Masoretic Text (Authoritative text of Hebrew Bible;, 6th. – 10 centuries AD).
C. Dead Sea Scrolls (portions of the Hebrew Bible; 2nd. Century BC – 1st century).

Pastors would find differences[22] in these texts and adapt the sermon to make the parishioners better understand the meaning and context of the Good Word. For example, one would find the following differences (underlined) in the above-mentioned texts:

> Version 1: "When the most high divided to the nations their inheritance, when he separated the sons of Adam, he set the bounds of the people according to the number of the <u>children of Israel.</u>

> Version 2: When the most-high divided to the nations their inheritance, when he separated the sons of Adam, he set the bounds of the people according to the number of the <u>sons of God.</u>

According to some Bible scholars, in the King James version, the word *hades* and *sheol* shouldn't have been used interchangeably since these are two different concepts.

Misinterpretations of some religious texts has also resulted in death. For example, jihadists have killed many innocent people after mis-interpreting the Qua'ran. Associated with this issue are the topics of morality and the idea of honor killing where misguided beliefs are directed by immoral and inhuman convictions. Any beliefs that lead to murder is of the devil and people should grow up and just dismiss these convictions. There are however, several deranged evil people who use religion to kill other people, when they personally don't believe in any type of religion, but still use it as an excuse to kill.

Please use your God-given brains and not rely on others' beliefs. That's why God gave you your own personal brain so that you don't follow other people and cults to death, prison and disgrace. During this period, for example, a Lori Vallow Daybell of Idaho US, was

---

[22] Dead Sea Scrolls: What apologists hope you'll never find out. Paulogia. 2023.

found guilty of killing two of her children because of her doomsday cult beliefs. **Life is much more precious than any belief.**

In regards to cults from a scriptural perspective; Pastor Micheal Grant (Sinless Perfection Teaching Doctrine. E.G. White) concluded that the four main cults are Jehovah's Witness, Mormons, Seventh Day Adventist and Christian Science. Some theologians disagree. Some cults however, have changed their fundamental principles through the years and may now be considered as Faiths or Churches instead of cults. Several churches have also changed some of their fundamental doctrines.

Old versions of the Bible exist that reflect the gradual corruption of newer translated versions as the translators fail to uphold the true meanings and proper context of the original Biblical text. Because several passages in the Qua'ran were borrowed from the Bible (according to Islam expert Jay Smith) people have used some of those borrowed passages **out of context.** For example, Former US President Obama when trying to explain that Islam is a peaceful religion, used Surah 5:32, which was borrowed from the Bible but has the incorrect context. This is why it's critical for Samoans to be very cautious of any new faith introduced into our islands. These islands have **more than enough churches to last a lifetime**. For any new church planning to come down to these islands, you are not welcomed – a suggestion by the prominent late Chairman of the National Council of Churches (NCC), Rev. Oka Fauolo, and I agree.

I have a simple and personal example on how a correct interpretation is crucial. I often tell people that I've been in jail for nearly 50 times – **and this is a fact.** Some of the people that I say this to, look at me with disgust, misinterpreting my statement. The fact is, one of my duties as a paralegal is to deliver documents to the prisoners and to accompany our attorneys to do translations/interpretations for their jailed clients. Similar episodes of misinterpretation are actually happening in the local churches and this is very unfortunate, and disgusting at the same time.

Some Biblical researchers have even used a mathematical approaches to prove their persuasion that the King James Version is the one true translation of the Word of God. One such scholar (Peterson, Brandon. Truth in Christ. 2023) used Deuteronomy 16: 11, "*...in the place which the Lord thy God hath chosen to **place** his name there*". The mathematical analysis emphasizes the importance of the number 1611, the year the King James Version was Published. Presentations[23] by one Biblical Researcher suggested his preferred versions of the English Bible:

1. King James Version
2. New King James Version
3. English Standard Version

Another related issue is the distortion of words. In a secular contest, the distortion of Martin Luther King Jr.'s words is believed to "enable more, not lessen, racial division within American Society", according to Hajar Yazdiha, Assistant Professor of Sociology, USC Domestic College of Letters, Arts and Science (1-14-23). Several scholars, civil rights activists and King's own children have long pointed out that the use of King's words, especially by right-wing conservatives, are too often, attempts to weaponize King's memory against the multicultural democracy of which he could only dream. Distortion of Biblical texts have become increasingly common not only in overseas countries but also in these islands, especially by rookie "pastors". These local rookie so-called pastors should be advised that local biblical scholars need to let them know about the requirements and qualifications needed to be a leader of a church as taught through 1 Timothy 3. These islands are made up of small village communities and people mostly know each other, so it is easy to determine that these rookies are not qualified. Maybe they can help me with my taro plantation instead?.

---

[23] Mathew Everhard: Trust these three translations. 2021

Readers should also realize that there are several other related texts known as "The Lost Books of the Bible" that contain various related information regarding the early church and Christianity, that are important to researchers and Bible College students. Some information found in other sources contain worthy texts not included in the modern texts like the King James version. For example, the Dead Sea Scrolls tells of a conversation between Noah and his father. *The Lost Books of the Bible* and the *Forgotten Books of Eden*, first issued in 1926, are known to be "the most popular collection of apocryphal and pseudepigraphal literature ever published". Some of the Lost Books of the Bible (some attributed to the Apostolic Fathers) include the Gospel of the Infancy of Jesus Christ, The Book of Enoch, The Lost Gospel of Peter, The Epistle of Jesus Christ and Abgarus King of Edessa, The Act of Paul and Thecla, Letter of Pilate to Herod and a few others. Contents of The Forgotten Books of Eden include: The Testament of the Twelve Patriarchs, The Psalms of Solomon; The Story of Ahikar and five others.[24] Information in these texts may provide some insights into some of the Biblical events, interpretations and concepts. Some of the so called "Lost Gospels" were written significantly later, in the 2nd and 3rd. centuries (BBC, Teach)and this would have counted against them. Radical interpretations can also be found in several books like "The Lost Gospel" by Simcha Jacobovici, who claimed that he had discovered the tomb of Jesus.

A recent related secular example is the Nonsensical translations of Alaska Native applications for typhoon damages recently (COVID-19 period) resulted in delays for Indians in getting government assistance. Native Indians in Alaska who mostly use *Yup'ik* and *Inupiaq* languages found bizarre phrases in the Federal Emergency Management Agency (FEMA) paperwork. An example in formal government information provided to the natives was:

---

[24] Goodspeed, Edgar J. Modern Apocrypha (Boston, Beacon Press, 1956).

"Your husband is a polar bear, skinny". Again, poor translation can affect people's lives.

The first translation of the English Bible to Samoan was done by a group of pastors and London Missionary Society (LMS) missionaries called the *Papa o Misi* (the rock of missionaries). The main change in a later translation is the replacement of the word *Ieova* with the word *Ali'i*. Translators themselves should trend lightly and carefully when dealing with scriptures so the faithful get the true intent and meaning of the Good Word. For example, Dr. Bart D. Ehrman, a New Testament scholar suggested that Mathew mis-translated the Hebrew word "virgin" (Isaiah 7:14) to Greek (Mathew 1 and 2).

During this period there were about 151 different English versions of the English Bible. Bible Scholars prefer different versions of the English Bible. Those who favor the King James Version have pointed to Texts like Gipp's Understandable History of the Bible; In Awe of Thy Word: Understanding the King James Bible; The Answer Book: A helpful Book for Christians by Samuel C. Gipp; The Language of the King James Bible by Gail Riplinger. Jehovah's Witness has its own Bible translation they call "The New Translation of Holy Scriptures" which predicts the end of the world. Armageddon was wrongly predicted at least five times in this version since 1914 and I personally attended some of the related events at *Lalovaea*, Samoa in the past. Why Samoans continue to believe this translation is beyond me.

Transliteration, however has been a concern to a few scholars of the Samoan language. A secular example to ponder: I have been personally saddened by a rookie local radio broadcaster, in American Samoa, who emerged in 2021 on 93.1 KHJ Radio and completely butchered the Samoan language through absurd translations and transliterations of several English words into the Samoan language. Of course, there are several English words (e.g., computer) which traditionally has no corresponding Samoan translation and a transliteration maybe appropriate. He should have trend lightly on this issue and use a **description or explanation** of the English word – a technique I sometimes use as a Licensed Translator. In the Rwanda language, there is no local word for architecture, so the people used an expression or description meaning: "expert in the creation of buildings". This local broadcaster used the transliteration "kulupu" for the word "group" and translated the word "show" to "siou" which sounded ridiculous and absurd. According to Samoan author and Samoan language authority Dr Semisi Ma'ia', this mistreatment of the Samoan language can me "lethal" and would eventually "hijacked" the original text. The American Samoa Community College (ASCC) Samoan Language Institute (SLI) started a useful related project a few years back where English words were offered to the public for their suggested translation. If I'm not mistaken, this project resulted in the now widely recognized translation of the word internet, to *upega tafa'ilagi*. ASCC, NUS and other entities should collaborate to continue this important task to ensure a sustainable, relevant and healthy language. I would be glad to assist as I had translated a few lines of the American Samoa Code Annotated when my father, the late Rev. Aitaoto Seiuli Fuimaono was the Chief Translator for the American Samoa *Fono* in the early 1970s.

In the past, various Christian writings were translated by several people from Hebrew into various languages including Arabic, Syriac, English, Latin, Greek, Aramaic, Samoan to name a few. Diverse interpretations and dissimilar perspectives of the Bible text has resulted in the multitude and assortment of churches, denominations, beliefs and the relatively new, **lack of faith**. Million others have their own personal "religion" and millions more don't careless. Other groups like the Hutterites, Mennonites, Hasidic Jews, Adas, Amish and others add a myriad of selfish interpretations (some to fit their preferred lifestyle) and doctrines to various communities around the world, adding to the variety and confusion towards religion and contradicting monotheism. The controversial idea by some scholars that Mary Magdalene was married to Jesus mostly came about due to the failure of these scholars to carefully examine the Greek word "*koinonos*" (Gospel of Philips) which variously translated in contemporary versions as partner, associate, comrade or companion. There are also various languages such as Pennsylvanian Dutch and Yidish (a mixture of English and German) that various communities use to explain their religious beliefs. For young Samoan readers, the word "*ekalesia*" in the Samoan Bible comes from the Greek word "Ekklesia" which in the New Testament mean "church", and refers to "a group of people who have been called out of the world and to God".

Which church has the correct interpretation of the Bible is anyone's guess!, since NO ONE can read the mind of GOD. It's like trying to explain to a 1-year-old the meaning of $E=mc^2$. Jesus' frequent employment of parables in the Gospels definitely has its own divine reason but has also contributed to the misperceptions and confusion that we now have to endure due to man's failure to utilize novel available resources and seeking guidance from the Holy Spirit.

In 2022 Pope Francis issued a new decree that required prior Vatican approval for bishops to erect new associations of the faithful, often the first step in the creation of a new apostolic society or institute of consecrate life. The root cause of this "breakaway" from mother churches in many countries including the Samoan Islands is the different interpretations of the Bible.

In these islands, other causes of breakaways include personal conflicts within churches, the refusal of some members of the clergy to serve under other pastors and sometimes it's the inappropriate meddling of the chiefs in church matters. In the Samoan islands, this phenomenon has increased in the past two decades and in some instances new churches have been formed under the leadership of locals with no or inadequate Biblical training. Depending solely on the help of the Holy Spirit is not sufficient because God has provided novel avenues and resources (e.g., the Internet) to get in-depth understanding of the Gospel. I have personally listened to several televised church services led by some of these Samoan rookie preachers and the superficial crux of their messages are sometimes misguided and demeaning due to their shallow and imprudent interpretations of the Bible. Due to their lack of biblical training and needed extended affiliation with the Holy Spirit, they have distorted the Bible text tremendously. Sermons from these un-trained clergy members are often trivial in nature and economical with the Biblical truths. This phenomenon has also resulted in the many offshoots (or splinter denominations) of new denominations springing up around these islands. This, however is not a new world-wide church experience. Judaism, for example has a few sects like Orthodox, Conservative, Reform and Reconstructionist. The Catholic Church also has a weird (research this on the internet) sect called *Opus Dei* whose members "seek personal Christian holiness and strive to imbue their ordinary work and society with Christian principles". Readers can also read the book "Der Weg" to get additional information on this sect.

A woman speaking on the local KHJ V103.1 radio religious program in American Samoa on February 16, 2023 said that there was a lot of problems facing the people of American Samoa because the churches have allowed bingo games to obtain money from the public. She pointed out a bingo game conducted by a Pavaia'i denomination. My research suggested that the poor lady never attended a Bible College, doesn't understand Hebrew nor Greek; mainly does her study using just the Samoan Bible and she was also a relatively new recruit to the Holy Spirit family. I've seen a few beautiful church halls and church buildings that were built using some funds from church bingo games and the pastors of these churches are intelligent graduates of Bible Colleges and Sprit-filled servants of God. I can personally take this woman to these pastors if she wants to discuss this issue with said pastors and if she wants to humiliate herself. Countries select their best athletes to represent them in Olympic Games and the Samoan Islands should also carefully select "qualified" people to preach the Good Word. The Gospel messages are from the omniscient All-knowing, Super intelligent, Super smart and Supernatural entity and therefore the messages from the Christian pulpit (and through radio and television) should reflect these heavenly attributes. Preaching and spreading the Gospel of Jesus Christ in the Samoan islands should be like the Olympics Games. Only the very best should be chosen to preach, because this is not an inconsequential undertaking. It's a vital and divine component of the Samoan indigenous human existence.

Here are a few hilarious secular examples of how this phenomenon can be very subjective in the world of commerce overseas. When Kentucky Fried Chicken (KFC) introduced its "finger lickin good" commercial into China during the 1980's, this was translated into Mandarin as "*eat your fingers off*". Rolls-Royce changed the name of one of its cars from Silver Mist to Silver Shadow when the company found out that "mist" translates to "excrement" in German. Be careful not to ask a waitress in Australia for a napkin because she will be happy to give you a diaper! A "take-out" in Pago Pago means a "take-away" in Apia restaurants. Amazon Company refer to warehouses as "fulfillment Centers".

The original translation of the Bible to the Samoan language should be the only text to be used during routine church services because that translation was done under the direction of the Holy Spirit. A later revised Samoan edition of the Bible was a project done when pastors, local and foreign Samoan Bible scholars dealt deeper into the translations and interpretations of the Word and concluded that a new translation was necessary. I view that new translation as a human scholarly and intellectual exercise and not a spiritually initiated project and therefore the newer version should be used only in Bible schools and personal studies and not in normal church services. This will also eliminate the muddle during the congregations group reading of the Bible during normal church services when parishioners read from different Samoan versions of the Bible. Remember, God is an **orderly** concept and His creation and craftsmanship was performed in an orderly manner. Prominent scientists have also surmised that some complex mathematics, Calculus and complicated logics (recently quantum mechanics) were involved in Genesis 1, where order is essential. Evidently, prominent mathematicians and scientists[25] have examined several examples of mathematical natural phenomena in nature and have pointed out that there are several patterns found in many petals and flowers that exhibit these patterns. Some exhibit the concept in the Fibonacci series and the concept of the constant "pi".π. All requiring order.

Related to this issue is the way most Pentecostal worship is done inside their churches. When I first joined the Assembly of God church, I felt disturbed by the sounds and movements of parishioners worshiping. People were very loud and sometimes run around the congregation and clapping loudly. I was a congregationalist for most of my young life, and this was new to me. This prevented me from connecting personally and having a quiet moment with the Holy Spirit. It was really hard for me to focus. I think the style of free worship is tolerable as long as one doesn't disturb other parishioners. But if one prefers a more somber and quiet type of worship, there are denominations that offer that style. One pastor brought up the idea of "The Law of first use", where the word worship was first mentioned in Genesis 22 and only two people were involved. There were no noise, no keyboards, no clapping and raising hands in praise and no microphones. Maybe these islands contemplate going back to the solemn, somber and lugubrious roots.

In addition to the messy "products" caused by the various translations and interpretations of the Bible, stories are sometimes told so many times throughout decades that readers believe in them and refused to look at the facts which sometimes contradict Bible stories and eye witness accounts. One English version of the Bible was the work of only one man that included his own personal translation and interpretation. The text is ridiculous. The Translation "Mirror

---

[25] Dr.Mario L "Is God a Mathematician" and "Our Mathematical World". By Dr. Max Tegmark

Bible" by Francois de Toit is seen by some Bible scholars as "anti-gospel", where the author has added his own personal interpretation and failed to utilized Greek correctly.[26]

Readers should keep in mind that the Bible doesn't conform to human reason or science, but to God's will. The various interpretations of the Good Word had also led to serious wrangles amongst doctrines, beliefs and practices between churches and within congregations. In 2022, 400 Methodist churches in Texas voted to leave their parent denomination, the United Methodist Church (UMC). This event followed the mass exodus of Methodist congregations in other states including, Louisiana, Georgia, Arkansas, Florida and North Carolina. The departing congregations joined the more conservative Global Methodist Church over concerns that the UMC has grown too liberal on key cultural issues like the LGBTQ.

Getting the correct interpretation of the Bible is necessary to avoid inappropriate use of millions of dollars which should have been used to feed the poor. For example, a US televangelist promoted a fundraising event that raised about 3 million dollars for the people of Israel (Eitan Var One for Israel Ministry 9/119/19) believing that he will be blessed (according to his interpretation of the Bible) since he is giving to Israel. Unfortunately, most if not all of the NGOS in Israel that received the funds from this event don't believe in Jesus Christ. An interesting statistic that was provided noted that 99.7% of Israelites don't believe in Jesus Christ around the time this donation was made. Another interesting statistic indicated that, at the end of 2022, Christians made up 1.9% of the Isarael population, 75.8% of the Christians in Israel are Arab Christians and Christians make up 6.9% of the Arab citizens of Israel.

Perspectives provided by the Replacement Theory, and the Dual Covenant Theology have added to current confusions in interpreting the Bible. Additionally, other theologies like the Process Theology which states that God is still developing and the Openness Theology

---

[26] Mike Winger "Literally the worst Bible translation I've ever Seen".

which believe that God can't know the future contribute more to the misperceptions. Established religions like Judaism also have a few different beliefs, sects and divisions within their churches, all adding to variety, contradictions and confusion.

In 2022 a court in India sentenced to death 38 people for a series of bomb blasts in 2008 that killed more than 50 people in Gujarat, which has a history of violent clashes between Hindus and Muslims; two different routes of religious beliefs.

The Duden dictionary, the leading dictionary of standard German changed its definition of Jew, or "Jude" in German, during this period, and caused an uproar in the country's Jewish community.

Christianity is not the only religion sector that have similar problems. In Islam, there are five different schools of Islamic laws; one Shia school and four Sunni schools (Shafi'i, Maliki, Hanbali and Hanafi) and the five schools differ in how they interpret the texts from which Sharia law is derived. This means that Sharia may look different in different places. Furthermore, culture and customs influence how these laws are interpreted. Sharia is Islam's legal system. It is derived from the Qua'ran, Islam's holy book, as well as the Sunnah and Hadith which documents the deeds and sayings of the Prophet Muhammad.

The Greek word translated as "inn" (*kataluma*) doesn't mean hotel in any kind of modern service industry kind of way. The problem isn't one of vocabulary. Luke knew the Greek word for a hostel or inn and uses it in his telling of the famous Parable of the Good Samaritan. That he uses a different language here means that Luke is describing something else here. The question is: what is a *kataluma*? In an article published in *New Testament Studies*, scholar Stephen Carlson argues that *kataluma* refers simply to space. The meaning, he writes, is that the place that he was staying did not have enough space to accommodate the soon-to-be small family. Typically, Joseph's family would have put them up in the guest quarters that formed part of the upper rooms of the house. These rooms would likely have included a sort of small guest bedroom. Unfortunately, as Luke stresses, there wasn't enough room for everyone upstairs, so they likely stayed downstairs in the main room of the house on the ground floor, **not a manger**. An interesting topic for discussion during Bible studies.

Bible translators may also entertain using new words being constantly added to the English language so the younger generation would better understand Biblical concepts and can better relate to its messages. For example, about 90+ new words were added to the Oxford Learners Dictionary sometime during this period.

I strongly believe that had the authors of the Gospels be given the technologies now available to modern authors, the scriptures would be different. During the editing phases of my books, I would often change/omit words a few times. Sometimes I would re-arrange the main structure of the text. A portion of the text that I had revised and checked before, seems to be fine at one time but after a few days, I would return with "a fresh pair of eyes "and change these again. The original text of the Bible would have been different if the authors of the Gospels were using laptops with the latest version of Word where they can edit and change the text as they wish. Editing hand-written scriptures would have been a monumental task that would be near to impossible during the penning of the Gospels. Joseph Smith **was believed** to pen more than 500 pages of the Book of Mormon in about three months. Seventy-two Jew scholars translated the books of the Old Testament, the Core Collection, to Greek in only 72 days. This was essential at that period since most of the Jews were speaking Greek when they returned from Persia. The result was the Septuagint. This perspective however, should be placed on the side as HIS divine plans can't be rationalized by miniature human minds. I believe this was HIS plan and HE knows what HE'S doing.

The Song of Solomon, which has several erotic dimensions, has been interpreted in various ways: It could be a portrayal of God's relationship with the people of Israel; Christ's love for his "bride", the Church; the soul's yearning for God; God's loving relationship with Jesus's mother, Mary; or human-to-human love; or lately, acceptable love between a husband and wife[27]. Interpretation is crucial for this book otherwise, Christians would be very uncomfortable studying this text.

In early 2023, the Codex Sasson, the earliest and most complete surviving manuscript of the Hebrew Bible ever discovered in book form, was offered up for sale for the first time in over three decades (Fox Business. Dumas, Breck. 2/17/23) The book is dated back to the late ninth to early 10th. century and is expected to bring as much as $50 million. Scholars should get copies of this book and delve into its contents to get as much information out of it as possible to further their understanding and proper interpretation of the Bible. There might also be problems and confusion in related matters. The Codex Sinaiticus has more than 3,000 corrections and additions according to some scholars and this could have resulted, for example, in the addition of text in the last part of the Gospel of Mark.

Archeology is very important to the study of the Bible. Recent presentations (Allen Parr BEAT 2023. Dr. Jeremiah Johnston). on the" Top 10 discoveries related to Bible" stories continue to strengthen the truth about the Bible. These discoveries[28] include the Galilee or Jesus boat, synagogues in the time of Jesus, Pool of Siloam, Jacobs Well, the Southern Steps of the Jewish Temple, the ossuary of Caiaphas, the Pilate Stone, and others. In 1986 the Turks announced the discovery (ARKDISCOVERY.com. "Truth is Christ)of Noah's Ark, using high-tech equipments by a group of scientists. Readers interested in this topic and the Greek Bible translation may also delve more into the Linear B Language. This is the oldest preserved form of Written Greek, a syllabic script used for writing Myceanaen Greek and was deciphered in 1952 by the genius British architect Michael Ventris. It is important to understand all Bible-associated languages. For example, the Arabic words "Et-tell" can mean a hill,

---

[27] The Conversation. Kaplan, Jonathan. The University of Texas at Austin. 2/10/23.

[28] Associates of Biblical Research. Brian W. biblical.archeaology.org, WBPH Studios.

a mount or even a ruin. Several Samoan Bible College graduates are familiar with this concept. Recent documentaries estimated that there are at least 5,500 Greek manuscripts of the New Testament, which may partially attributed to this concept. There is also a related archeological discovery that suggested that the number 666 – allegedly the mark of the beast- should be 616. This was announced by a scholar after two years of painstaking work examining a very old manuscript. (Demar, Gary (2005) The American Vision Ministery) Examinations of old scrolls from various monasteries (e.g., Hemis Monastery) in several countries revealed that Jesus (or Issa- Yeshua according Buddhists Monks) might have been living in India, Tibet and Persia. This period has often been referred to by several writers as "the lost years of Jesus". More discoveries will be revealed in the future as scientists continue to search for more evidences regarding the Bible. In fact, a scientist (Business Insider. Iemeyer, Kenneth. 5/22/23) announced in May 2023 through a peer-reviewed academic journal published by the Cambridge University Press, that he found a chapter of the Bible hidden for more than 1,500 years. He and his associates used ultraviolet photography to spot the text, which was hidden underneath multiple edits and would offer a gateway to understanding the earliest phases of the textual evolution of the Bible.

When I attended the LMS Sunday schools in the past, I always thought that God created different things each day from Day 1 to Day 6, but this is not the case according to some scholars. If one reads the Bible carefully, God didn't create anything on Day 2. The waters were already on Earth (God already created this, the author may have missed this) and he just needed to divide the waters. Some scholars believe that there were civilizations before humans. Adam was however, according to Genesis, the first human. According to one Biblical scholar, if one utilizes the Old Hebrew language, then the correct translation in Genesis should be: "In the beginning, the **Gods** (plural) created the heavens and earth". This should be good news for those subscribing to the Trinity conviction.

The question on how old is planet Earth has two main proposals.

1. About 13.7 billion years old with no intelligent design or creator.
2. Created in 6 days (Genesis).

While determining the age of the rocks by using various samples and various methods, the results seem to provide different numbers that sometimes differ by millions of years according to Geologist Andrew Snelling Ph.D. So the saga continues.

People can sometimes simply use the wrong words in their writings and so the prospect users will get the wrong impression of what actually occurred, or what the writer meant to impart. For example, a former President in one of his tweets in early 2023 used the word "indicated" when he meant to use the word "indicted" – he was later indicted. At the same time, a Legal consultant mentioned that the word "arrest photo" should be used instead of "mug shot". We all make honest mistakes sometimes.

Digging into what some old texts writers might have intended to impart sometimes reveals amusing reckonings. For example, the US Constitution mentions "we the people". After some research, I thought this phrase was intended to exclude women, slaves and Native American, and several other scholars have the same opinion. Additionally, women were excluded from politics in historical Athens, a birthplace of democracy[29].

A scientist recently discovered a lost fragment representing one of the earliest translations of the Gospels[30]. Ultraviolet photography was used to look at past layers of text to find the "new ancient translation" and the text is one of only four examples of the Old Syriac translation. The long-hidden chapter is an interpretation of Mathew Chapter 12, and was originally translated as part of what

---

[29] David Wengrow. TED. "A new understanding of Human History and the Roots of Inequality.

[30] Popular Mechanics, Newcomb, Tim. 4/10/23

are known as the Old Syriac translations about 1,500 years ago. The news of the discovery noted that "while fragments of New Testament text date back to the original writings from the 3rd. century, the oldest known surviving complete manuscript of the New Testament is the Greek Codex Sinaiticus, dated to the 6th Century. God's message from Jebel Musa (mount Sinai) continue to bring controversies for Bible scholars.

Samoan parishioners should be aware that erroneous interpretation of the Bible can be dangerous (e.g. disallowing blood transfusion needed for emergency medical operations, or stoning someone to death) and may also lead to illegal acts. For example, a Wisconsin teen was sentenced to 20 years in prison during this period for the 2018 death of his 7-year-old relative, who prosecutors say was beaten and tortured (Associated Press. P. M Burke). The teen's father had instructed him to punish his relative because he didn't know the first 13 verses of the Bible. People of the Samoan Islands should distance themselves from evil entities that place their Bible interpretations before human life. There is absolutely no correct interpretation of the Word of God as no man can read the mind of the Almighty. Human life is much more precious to Him than any questionable interpretation of his Word. This is not an easy subject; even Apostle Paul and Peter had their own different interpretation of the Christian message.

US federal Grantors shouldn't waste their funding money to those researchers trying to research the origin of all different languages in the world. The explanation is in Genesis 11. Readers interested in this topic as it relates to history and the classics may also be interested to know that there is now a new translation of the epic Iliad, a story of honor, war and reputation, by Emily Wilson. Hundreds of dollars were wasted on grants for some researchers who were trying to locate the Garden of Eden in the past decade. This is an impossible task according to one anthropologist.

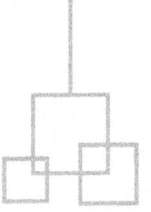

Celebrations and Special Days

C hristmas Holidays have always been Special Days for the American Samoa Government and the American Samoa Council on Arts Culture and Humanities where many Government Departments and various groups provide carols and other forms of entertainment for several days. In 2022, this program was held at the Veterans Memorial Monument in Tafuna.

The Day of Atonement (Yom Kippur) is known as the holiest Jewish holiday and it comes ten days after the Jewish New Year (Rosh Hashanah). Jewish people are supposed to spend quality time doing self-introspection and expressing regret or remorse to those that may have hurt. Jewish people often refrain from work and spend most of the day at religious services. There were no Jews in American Samoa during this period.

If one passes by the High Court of American Samoa on Wednesday(s), one would normally see colorful attires of people crowding the patio. This is the day assigned to weddings, and the couples' families and their friends often come to the High Court to take photos on this special day. Then on the following day, the court sometimes deal with divorce cases!.

Ululation should be permitted during funeral services. As I mentioned in my previous Book, Celebration of life during funerals is a relatively new concept to the Congregational Churches who first arrived in the Samoan islands bringing Christianity. The relatively new Pentecostal Churches introduced the concept of "celebrating the life of the deceased". Grieving and wailing are natural to humans and also animals (refer to my story about a family of elephants in my previous book) and should be permitted during funeral services together with the relatively new "celebration" view. "Grief is something you will have to carry with you for the rest of your life, in spite of what certain people or culture wants us to believe" - this was Lisa Marie Presley's final message on Instagram (1/13/23). According to a psychiatrist at the local LBJ Hospital, people who have lost a loved one should be left or allow to grieve. They need time to process the situation. Relatives and friends should just listen. In this type of situation, the best method of communication is just listening and not expect or prod the grieving people to clap their hands and praise!.

A *faletua* of the *"Taulogologo I le upu a le Atua"* church from Leulumoega Tuai won a brand-new Ford Ranger courtesy of a Digicel Samoa promotion during this period. A very special day for this *faletua*.

In 2022 the village of *Sapapalii* held celebrations on August 24, to commemorate 192 years since Christianity first arrived on the islands of Samoa. The history of the event described the arrival of John Williams and his wife Mary in 1830 on their ship the *Messenger of Peace* – a ship which was supposedly built by Williams with local materials in Tahiti. The passengers on this trip included a Samoan couple *Fauea* and his wife *Puaseisei* (a native of *Safune*). The passengers later met with Malietoa Vainuupo, then king of Samoa. Williams later noted that Samoa was the only country in the pacific where Christianity was accepted without war. King Malietoa later declared that the traditional monarch and kingship would stop after his death and that Samoans should turn to Christianity as their way of life. The LMS church was established and it quickly sent out missionaries to other pacific islands as early as 1839.

The first Sunday of September is commemorated each year by the Congregational Christian Church of American Samoa and

Samoa as *"Aso Sā ole Tala Lelei"* or "Gospel Sunday". It marks the arrival of Christianity, August 24, 1830 at the village of *Sapapālii, Savaii*. This was the dawn of Christianity in Samoa. Missionary John Williams, of the London Missionary Society, brought the Good News to Samoa on the boat, "Messenger of Peace". September 5, 2021 marks 191-years, of the arrival of the church in Samoa. The village of Leone, where John Williams and LMS missionaries arrived on October 12, 1832 was to commemorate 190-years in 2022. A commemorative structure of John Williams and his arrival is located in front of the CCCS church in Apia.

John Williams Monument

These historical events are true and factual as oppose to the political party HRPP's claim that "the darkest day in the political history of Samoa was the day when they were locked out of the parliament building". HRPP also advocated that this event should be memorialized so that future generations can know the truth. This claim is **absolutely false**, incorrect and untruthful and I feel sorry for future generations if they take these fictional stories as historical fact. Now let me correct this false assurance because I also want future generations to know the truth and hope they memorialize the actuality of Samoa's history. Here is the absolute truth. **The darkest days in Samoa's political history are the days Tamasese Lealofi was killed in Apia in 1929 under the New Zealand administration and the day HRPP locked out the Chief Justice and others from the Parliament House (*Maota Fono*) which consequently resulted in the swearing in of new parliamentarians under a tent.** HRPP claimed that the swearing in of the new members of Parliament was illegal and unconstitutional, but it was their actions that caused the swearing in to take place under the tent. Ridiculous, erroneous and deceptive claims. Why is this correction important to church-related events in these islands?. Well, many years ago, the renowned female warrior *Nafanua* advised then King Malietoa that the next administration of these islands would emerge from heaven. The prediction was spot-on; Christianity later arrived in 1830 and the churches have now become the greatest influence in the islands' political and cultural administration.

Yom Kippur (translate to "a day of Atonement") is considered the holiest day of the year in Judaism and is observed from sundown Wednesday to sundown Thursday and follows the Jewish New Year (USA Today 9/13/21 Carly M.). The customary greeting for this event is "G'mar chatima tova" and translates to "May you be sealed in the Book of Life".

In August 2022, many churches in Upolu marked Father's Day with songs, feasts and laughter on that special Sunday. Many family members flew in to join the celebrations and there were a variety of activities provided for this occasion. The men of the Methodist Church in *Satitoa* wore green ties and had flowers on their pockets. The children of the *EFKS Mutiatele* were served ice cream. It seems most people on these islands enjoy the usual huge meals but this is definitely not a good day for the local pigs and chickens.

American Samoa's Governor and Congresswomen offered official White Sunday (*Aso Sa Pa'epa'e*) messages to the people of the territory in White Sunday 2022. Samoan Christian diaspora in many parts of the world also celebrate this special day often with an elaborate *to'ana'* (Children Sunday brunch/feast)and children wearing all-white attires to church where they recite Bible verses. In American Samoa, the following Monday is a federal and local holiday.

In early 2023 the Samoa Government continued its customary Week of Prayer and Fasting which started in 1990 after extreme hurricanes *Ofa* (1990) and *Valelia* (1991) seriously impacted the country. Around this period, American Samoa's Congresswomen Amata attended the National Prayer Breakfast at the Capitol. This is an annual bipartisan tradition since 1953 when President Eisenhower attended at the suggestion of evangelist Billy Graham.

Christmas Eve 2023 was marked by continuous fighting between Hamas and Israel Forces. May be this is because December 25 is actually not the correct birthday of Jesus Christ. There is also the indifference of some people towards Halloween. Personally I would continue to allow my kids to enjoy these special days, and all other holidays for that matter, and forget about adults' interpretations regarding these days. The smiles on the faces of the children far outweighs any opinion and interpretaion on any matter.

The coronation of King Charles of Great Britain in May 2023, while I was writing this book, should be an important event to the local Christian community, as this was the nation that brought Christianity to these islands. It was important enough that the Head of State of Samoa travelled to Britain to represent Samoa on this monumental occasion. American Samoa was represented by US officials.

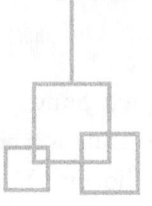

Language

T he Samoan language is the most important aspect of the Samoan Culture. It was the London Missionary Society (LMS) missionaries that helped formalize it, and proliferated it mainly by utilizing their *Pi Tautau* (Samoan alphabet chart) in villages *"Aoga a Faifeau"* – LMS pastors' school. This is a very significant contribution by the western Christian community to the Samoan language and culture. Readers who are interested in this topic can research papers by my friend Lonise Sapolu. Additionally, a few foreign scholars like Dr. Augustine Kramer had documented language-related information in their books. There are also dictionaries of the Samoan language available. Two prominent examples are the "Samoan Dictionary, Samoan-English, English-Samoan". (G.B.Milner 1966) and the old "Samoan Dictionary: English and Samoan, and Samoan and English; with a short Grammar of the Samoan Dialect" (George Pratt. London Missionary Press. 1862). A new dictionary is also now available: "Tusi Upu Samoa Volume 1. Samoan to English" by Papali'i Dr. Semisi Maia'i. Modern language experts like Papali'i Dr. Semisi Ma'ia'i, Tuiatua Efi and others have continued to document, critique and add depth to the language. The American Samoa Department of Education, during this period also added the ASIA program for students to learn the Samoan language during school breaks. During this period, Educator Muliagatele Vavao Fetui of the University of Auckland graduated with a PhD in Pacific Studies with his thesis written in Samoan. The first PhD thesis written in Samoan was by Taiao Matiu Tautunu, a National University of Samoa (NUS) Lecturer.

In December 15, 1976, following the admission of Samoa to the United Nations, then Deputy Head of State of Samoa, Tupua Tamasese Lealofi IV, addressed the Assembly in the Samoan language.

The General Secretary of the Congregational Christian Church of Samoa (CCCS.), Reverend Vavatau Taufao, prefer the Government address nationally important issues in both English and Samoan. This came to light when the Director General of Health refused an English response to a question by a reporter. A ridiculous defend of this government policy was printed in the government newspaper in November but was printed in English! Conducting official legislative business in Samoan is different from delivering information to a populace who speak English, Korean, Chinese, Samoan, but who mostly understand English. Information for the public should be in both English and Samoan. The delinquency with the HRRP administration's alleged policy of answering questions only in the Samoan language also came up when then Prime Minister said in 2021 that one of the newspaper reporter's mis-interpreted one of his remarks. This was government's fault. Formal public remarks shouldn't be mis-interpreted if the remarks are made in both Samoan and English. Using both languages and then quoting the source would solve similar problems.

Around this period, the Church of England launched a "project of gendered language"[31], in its effort to solve the problem of using male imagery to describe God. The project re-examined the use of "Our Father" versus "Our Mother" as Christians all over the world, for centuries, have used male pronouns to describe and communicate with God. The notion sounds ludicrous but it can be a reasonable question. One staunch Catholic said that he once believed that God

---

[31] The Telegraph. Lewis, Jemima. 2/8/23

preferred the company of bearded men. I also observed that maybe the Trinity view may also suggest to pray "TO THEM". This issue may not be of a gender binary assessment but more of a linguistic check.

Some scholars[32] have stated that the most important attribute of human tribes is their language. Of the other species, when they die, the information also die with them. Humans on the other hand, have the advantage of language, where they pass down information to future generations enabling them to dominate and live successfully on Earth. Dr. David Christian called this phenomenon "collective learning".

---

[32] TED. Dr. David Christian.

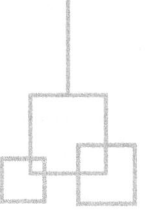

Mental Health

Mental Health is a person's condition with regard to their psychological and emotional well-being and how the pressure from various circumstances affects the person's personality, perspectives and actions. The impact of COVID-19 on the economy of the local community and its ripple effects on the youths and many adults was noticeable in the Samoan islands, especially in the church community.

According to a RNZ Pacific Journalist, Samoan Church leaders have been trained to offer mental health support to the members of their congregation. The New Zealand Mental Health Survey (2006), found 47 percent of Pacific people had experienced a mental disorder at some stage during their lifetime, compared with 39.5 percent of the overall New Zealand population. Pacific people were also less likely to make a mental health visit to a health service. The *Ta'iala Mo le Ola Manuia Mental Health Project in 2019* was initiated due to the growing awareness of mental health issues within the Pacific community. Pasifika-led non-governmental organization *Le Va* provides mental health, addiction, disability, public and general health, suicide prevention, education and in sport.

More than ten years ago, an Assembly of God (AOG) pastor in one of his sermons concluded that the people who are seen sitting in front of local stores begging for food and money are to be blamed for their own predicament and situation. He said that it was their personal decisions to become beggars. I don't think this pastor should ever preach the Word of God!. Did he ever consider mental illness? which is indeed the root cause of the demeanor of most of these unfortunate people?. Ill-informed people shouldn't stand behind any church pulpit, especially those that unexpectedly become richer with money from the church but have been poor all their lives. I can say this because I personally know such pastor.

In early 2023, it was announced that Japan has funded a new Goshen Trust mental health facility in Samoa. The funds were for the construction of a two-storey building to provide safe and proper health services for clients and cater for a larger intake of patients.

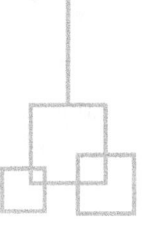

Diverse Abuse and Domestic Violence

This topic is important to local churches as the community need the churches' help, in partnership with the governments and NGOs, to make imperative headway into resolving related issues. Other countries in the pacific area are also experiencing this problem. New Caledonia crime statistics showed that domestic violence in 2021 was seven times higher than in mainland France (RNZ, 3/22/22). A nation-wide study conducted by the National Human Right Institution (NHRI) in 2018 confirmed the prevalence of violence within communities. When a young mother who was also pregnant was assaulted at Toamua, Samoa in May 2022 on Mother's Day, and in front of her three-year old daughter, it reflects a sinister infirmity that must be dealt with immediately, and needs the help of the churches. The churches have participated in related programs in the past decade and hopefully will continue to assist with this problem. An interesting note that needs to be mentioned is that males are not the only preparators, sometimes it's the female partners that are the abusers. For example, police responded to a domestic violence disturbance in Malaeimi, American Samoa in July 2022 and arrested a woman for domestic violence. This issue is also not new to some churches as well documented cases have reached US courts several times in the past two decades. Churches therefore may use their experience to assist the authorities curb this serious issue, but should refrain from utilizing their own processes since their people are not formally trained, nor Board Certified on this issue. Furthermore, this is not the calling of the church. Recent research suggests that many conflicts within families especially in marriages, evolve from physical problems of the human brain. Using brain scans, doctors have determined that the brains of some of the perpetrators show some abnormal scans. In one case they tracked down the workplace of one patient and discovered that he worked in a furniture shop where he was exposed to paints, polish and other related chemicals eight hours daily. The patient was asked to find another job and within six months his brain scan showed improvements. Consequently, there were no more conflicts within this marriage and the family was happy that the diagnosis went straight to the root problem -the physical human brain. In most of these types of cases and conflicts, **I strongly recommend a brain scan, and assessments of workplaces and extramarital relationships, instead of church counselling.** Patients planning to see a psychiatrist should also request a brain scan before wasting thousands of dollars on drugs and sessions that always treat the symptoms but completely miss the root causes. Contemporary research showed that a lot of psychiatric drugs shouldn't be taken for long periods and some of these drugs do more harm than good.[33] I personally know a case where a local pastor counselled a couple who were planning to get divorced. After the counselling the couple made up and were back together. After a few months, they decided to get divorced again. For ANY problem, I always recommend to determine the root cause then go from there. Spiritual approaches should be the last resort, because humans are first flesh then spirit later (Genesis).

There are churches e.g., Catholic, Jehovars Witness and Mormon[34], that have their own processes that deal with this issue. Recent researches and investigations have revealed the overwhelming

---

[33] Dr. James Davies (Ph. D Oxon) University of Rochampton
[34] BISBEE, Ariz (AP) Michael Rezendes.

inept nature of these processes. Clergy members trying to treat mentally-ill people had also resulted in death. For example, in 2016, a Mr. Murray killed Father Rene Roberra in Georgia after the kind priest helped him out for several years. This is what happen when the churches overstep their authority and capabilities and has also resulted in numerous cover-ups and consequently legal problems. This issue should be left to trained professionals within governments and various non-profits organizations. When millions of dollars have been paid to related victims (more cases are pending in court) and knowing that these funds were **from church offerings**, I suggest that the ruling members of the clergy in the churches allegedly involved, should all be fired. Human Rights abuses have been taken up by the Pacific Conference of Churches during this period against the Indonesian Government and suggested the positive influence of churches in communities, but not towards treating mentally challenged people.

According to Reuters, January 11, 2021, thousands of infants died in Irish homes for unmarried mothers and their offspring mostly run by the Catholic Church from the 1920s to the 1990s according to a formal inquiry. It found an "appalling" mortality rate that reflected brutal living conditions. The report covered 18 so-called "Mother and Baby Homes" where for decades young pregnant women were hidden from society, is the latest in a series of government-commissioned papers that have laid bare some of the Catholic Church's alleged worst abuses. It is comforting that the church has asked for forgiveness in these types of oppression. For example, in 2022 and in regards to Canada's residential schools, Pope Francis said: "I am deeply sorry. Sorry for the ways in which, regrettably, many Christians supported the colonizing mentality of the powers that oppressed the indigenous people. I am sorry. I ask forgiveness". This is a great man of God and the church should listen to him and practice the art of asking for forgiveness. Other churches should also follow his example.

RNZ reported in December 2020 that a Pasifika survivor is

calling for the victims in New Zealand's Pacific community who have been abused in care to seek Royal Commission of Inquiry for guidance and support. Frances Tagaloa, aged 52, spoke at the faith-based redress hearing in Auckland about sexual and emotional abuse she experienced in the Catholic church in the 1970s. Readers should be reminded that similar events also occur in other denominations. If you disagree, let me know and Ill dig up related information, but you have to pay.

In December 2021, the Samoa Police Service expressed concern at the rising number of domestic violence in the country in the last three years. TV1Samoa reported Assistant Police Commissioner Auapa'au Logoitino Filipo raised the concern with the media after the death of a young mother of six, allegedly at the hands of her husband early that week. Auapa'au said the woman was between 30 and 40 years old and died of stab wounds from a knife allegedly inflicted by her husband who is in police custody. The assistant commissioner said that 878 domestic violence cases were reported in 2019 and 964 were reported in 2020. So far, their office has received 628 complaints of domestic violence. For the latest domestic violence fatality, the Assistant Commissioner said that as the investigations are continuing, he cannot say what the husband has been charged with.

According to Samoa News (6/16/2021), a prominent California minister with ties to Samoa has been accused of embezzling more than $100,000 from a disabled veterans who lived in a filthy building next to his church. The good pastor, who heads the Second Congregational Church of Samoa in Long Beach, has allegedly been charged with felony and massive theft and theft from his dependents. The *California News Times* newspaper reported the story. According to a 14-page affidavit filed by the state prosecutor, the pastor was accused of embezzling thousands of dollars from the disabled veteran Peter Campbell, who was found living in a filthy building next to the church. Campbell, a veteran in his 60s, suffers from schizophrenia. The *Long Beach Post* also reported that

Campbell received nearly $ 3,000 a month from the US Department of Veterans Affairs. In 2016, the minister applied for a power of attorney to manage money on behalf of Campbell. Over the next four years, the pastor was alleged to have received nearly $170,000 and has been accused of spending much of that money on "suspicious purchases" of clothing, jewelry and audio equipment. According to the *Daily Mail newspaper,* Campbell, who once served in the US Air Force, left his sister's home in North Carolina in the early 2000s. He suffered from mental health problems and soon stopped contacting his family. Campbell's relatives spent years looking for him for fear of his health and safety. Finally, a Google search for his name in 2017 led them to the church. The minister is well known in the Long Beach community and is known for working with the homeless. Maybe he has a different story to tell.

RNZ Pacific Journalist reported during this period that a survivor of abuse, Frances Tagaloa, in a faith-based institution is appalled at what she calls the Catholic church's mishandling of victims' stories of abuse. Frances said that "It seems that the Catholic church is struggling to help victim survivors and are mishandling the situation.". The comments followed the recent Abuse in Care Royal Commission of Inquiry Pacific investigation hearing in Auckland.

In 2022, the Director and Founder of *Pasefika Mana Samoa Social Service Trust* told the Samoa Observer that the abuse of children was festering in Samoa. The NGO was setup to help abused children and the Director believes that related government ministries generally look down on NGOs, but she needs everyone involved to work together (Samoa Observer 19-2-2022 L. Hald)

Too often, current political issues addressed by Christians overseas gets reduced to a few topics like same–sex marriage and abortion. This dynamic seems odd, since Jesus never talked about those topics. The Gospels, however indicate Jesus talking about economic justice, improving Social Programs and fixing tax issues. Taxing local pastors was a prominent issue towards the end of

HRRP's rule and their indifference towards this issue resulted in its sudden collapse.

Traditionally, the physical abuse of children was not a very serious issue to most Samoan parents, who often discipline their children by "*sasa*" – caning or smacking the kids' behind with a broomstick. The arrival of Christianity in 1830 brought foreign but similar concepts like that in Proverb 22:15 which was interpreted by many Samoan parents in a biased way and utilized it as a justification to *sasa* their children. Later when the rule of law was introduced in the 1960s through the newly established constitution, the issue then had an avenue to be brought to the attention of the community and the government. Government, NGOs and the church in later years started to promote this issue and took steps to safeguard the children and women. The recent (May 2023) termination of a school teacher of the Wesley College in Faleula, Samoa, for using the *sasa* on a student is an indication that the Samoan community and the government are serious in disallowing this practice to continue. It is interesting to note that a few pastors and adults I interviewed in the past decade and recently, still believe in the *sasa*. A few years ago, I attended a relative's funeral and the church service at *Fugalei*, Samoa was delayed for about thirty minutes because the pastor who was supposed to lead the service was making a presentation at a workshop discussing the abuse of children. When the pastor arrived and the funeral service got on the way, the testimonies of the deceased's man's children provided very interesting stories of how abusive their father was. The message that the children was underscoring is that it was their father's abusive behavior that consequently resulted in every one of them being very successful in life. In fact, these children held high positions in government and private businesses, and one is a Pastor. After the children's testimonies, the pastor said that he had just returned from a workshop and he spoke against the abuse of children, but now after hearing the testimonies from the deceased's children, he now changes his mind and suggest that parents in the village continue to *sasa* their children, but in a mild physical manner.

The church erupted with laughter but the pastor seems serious and I was unsettled in my thoughts.

In American Samoa, the US Navy came in during the early 1900s with their own interpretation of the US laws. Traditionally, related matters were not normally brought into the "legal authorities" (Council of Chiefs) attention. This group governs the way justice is practiced in the villages and from the viewpoint of western laws, this is often "unequal justice". Witnesses are often not called to testify and details of events are usually not properly examined. From experience and my first-hand interviews for this book, many children that were well disciplined using the *sasa*, usually end up well in society (I also discussed this issue in my previous book). Of course, there are those children that basically have mental issues of varying degrees, and *sasa* would only make matters worse and there would obviously be no expected improvements in their lives. The issue of domestic violence was not usually included in the sermons of Congregationalists until the past two decades. A pastor of a Manua EFKAS in 2022 during his sermon on KVZK television said that "domestic violence has to stop".

During this period, a Samoan man was jailed in New Zealand for human trafficking and treating people from Samoa like slaves. Joseph Auga Matamata was jailed by the Napier High Court in 2020 (Samoa Observer. 2/5/22). General abuse is not a western concept, it is also present in the Samoan islands with the treatments of foreign nationals by their immigration sponsors. In 2022, three US states banned slavery but one (Louisiana) voted to keep it. This was about 160 years after the Emancipation Proclamation, the order that freed slaves in states rebelling against the Union during the Civil War. A few churches are also the culprits. For example, a case involving the Faith and Life Changing Ministries in New Zealand. This ministry, under Pasto Alofa T, tried to deport a Samoan man after the man complained of not getting paid for several months. Other countries like Britain continue to have domestic slaves while more than a million migrant slaves exist in Thailand's fishery industry during this period[35].

The Samoa Vitim Support Group (SVSG) with its partner Men and Boys Against Violence Project had a consultation with the *Avao* village on the contributing factors and possible solutions to violence in 2022. A Community Guide was developed that was based on the theme "*O Atamua o le Taeao o Mataitusi Auro Fa'aalofiafi o le Tusi Paia e Tafaesea ai Sauaga*" (The new morning of the golden burning letters of the Holy Bible eliminates violence). The SVSG had entered into an agreement with the government in 2022 whereby the government will provide support to deliver programs to eliminate violence against women and girls.

The village of Vailuutai in Upolu completed a 2-day workshop in September 2022, where a Community Guide was drawn up as a tool to eliminate violence against women and girls.

In September 2022, Samoa's Ministry of Women Community and Social Development received the prestigious award of the Global Spotlight Initiative from the "Leaving No One Behind" project. The event took place at the Paradisus Conference Hall in Cancum, Mexico. According to UN Resident Coordinator in Samoa, "in, many ways, the award gives voice to the painful silence surrounding victims of domestic violence, women and children who are denied the basic right to a dignified life".

The SVSG Samoa Victim Support Group humanitarian organization held a two-day workshop in partnership with the village of *Lalomanu* in late 2022 to address and eliminate violence against women and girls. *Lalomanu* is one of a few villages with only one church (CCCS). According to village chiefs, the village's two Meeting Houses, *Olotavale and Mateilonanuu*, are the two formal places where important matters, including abuse of women and girls, are discussed. After the arrival of Christianity, a third meeting place was established to discuss these matters. This was the CCCS church

---

[35] ViceAsia "Modern day Slaves of Thailand".

building. A related Community Guide was drafted utilizing input from all sectors of the village community.

In October 2022, Star Kist company held a lighting ceremony to remember the victims of domestic and family violence. October is the Domestic Violence Awareness month in American Samoa and the Department of Human and Social Services annually hold a candle light ceremony to remember the victims. Participants all wear purple and Star Kist and government Department of Health and Human Services (DHSS) employees lit five candles at the company's Satala cannery.

In Samoa, a government-funded project aimed at educating the public about violence against women and girls was implemented in *Iva Savaii* in 2022. (My late grandmother Alieta Magele was from *Iva*). The program advocated young men's economic empowerment as one of the preventive approaches to this issue (I personally and emphatically agree with this approach). Tools and equipment valued at around $5K tala were presented to the untitled men of the community.

The alleged confession by a former Samoa member of parliament in 2023 that he had abused women and girls indicated that this problem is real, serious and has infected every level of the Samoan community. His apology, admission that he has an illness and his beckon for medical intervention, heightened the need to address this problem from a variety of approaches including the participation of churches and the medical community. In fact, the offender later became a pastor for the SDA church according to Samoa News media. A statement from the Samoa Police around this time indicated that there was an increase in rape and sexual assault cases that had gone to trial.

Biological fathers abusing their biological daughters is very rare in these islands but in February 2023, a father was handed a prison sentence of 18 years for raping his 14-year-old daughter. Church influence and religious impacts on this matter can only go so far. A 2018 Report by Samoa's National Human Rights Institute found that more than 80% of women in Samoa are in abusive relationships and reported being punched, kicked or beaten with an object by their Intimate Partner. A lesser percentage report being strangled and one in ten women report that they have been raped in their lifetime by a family member or someone known to them. Samoa Police has a special phone number (22222) to contact in these types of cases.

In April 2023, a Pennsylvania grand jury accused nine men with connections to the Jehovah's Witness, of child sexual abuse in what some consider the "nation's most comprehensive investigation yet into the abuse within the faith". A similar grand jury investigation into child sexual abuse by Catholic priests culminated in a lengthy 2018 report that concluded hundreds of priests had abused children in Pennsylvania over seven decades and church officials had covered it up. A similar report was also recently issued in Maryland. My problem with this issue is that church elders have been treating this issue as a sin and not a crime. My friends, child sexual abuse is BOTH a sin and a crime. These are not the only churches that these acts have been attributed to. In fact, most if not all churches are guilty of these acts in various degrees, and this is a worldwide problem (readers can do their own research on the internet to verify my claim). In fact, during this period, the National Council of Christian Churches in collaboration with the Pacific Conference of Churches held a conference in Apia that discussed "attaining a biblical approach to end gender-based violence". The conference had two objectives: First, to introduce gender justice theology, which is a biblical approach that aims to end violence against women and girls, children, and people with disabilities, and two, to introduce the gender status card which is the result of the Pacific Conference of Churches General Assembly in Auckland in 2018. The card is a self- assessment tool that churches can use to determine and assess where they are in relation to addressing gender issues in churches.

An interesting related decision was made by the village of Vailu'utai during this period. The Village Council decided not to allow women who hold chiefly titles to participate in Village Council meetings. The Council noted that this was out of respect because sometimes the discussions were too hostile. The Village Council however pointed out that it honors the women's contributions to the village but at the same time it doesn't want to expose the ladies to hostile discussions, especially when discussion include sexual matters. The Samoan culture is very specific on the issue of women and girls. Men must treat the ladies (*feagaiga*) of the family with uttermost respect. This village was one of seven villages working with the Samoa Victim Support Group with government financing towards eliminating violence against women and girls.

In 2023, American Samoa's Victims Advocate Mrs. Ipu Avegalio Lefiti received the National Award for her work with victims of sexual and domestic violence. This was the Visionary Voice Awards presented by the National Sexual Violence Resource Center. Each year, state, territory and tribal coalitions select an outstanding individual to nominate for the awards. The American Samoa Alliance Against Domestic and Sexual Violence nominated Mrs. Lefiti that year. It was interesting to me that she mentioned systematic coverups, local attitudes and the sense of silence as hinderances to her work and that almost every door was shut in her face as she advocates for assistances for survivors. I totally agree with her since I assisted in a couple of her cases.

Local organizations that are interested in getting an in-depth view on how to work with "Samoan victims, perpetrators and *aiga* affected by family violence" can read the paper titled: "*O le Tofa Mamao* – A Samoan Conceptual Framework for addressing family violence", available on the Web.

A former local victim of abuse that was housed at the SVSG for about six years, returned to Samoa in 2023 after 18 years living overseas. She was adopted by a family in Utah from the SVSG compound and had since earned a degree in Criminal Justice. She spent time with current victims at the Campus of Hope sharing her story and sharing quality time with victim-residents at the SVSG campus.

During this period, the National Human Rights Institution inquiry found that nine in ten Samoan women have experienced emotional of physical violence at the hands of family members. The inquiry also noted that six out of ten had experienced intimate partner violence; one in five women said they have been raped by a family member or someone they know, and that 33 per cent of women who are raped contemplate suicide while 13 per cent attempted suicide. The survey offered two phone numbers that victims can call: 22222 or 800-7874.

Police in Samoa continued their investigation during this period into an alleged attempted rape of a 15-year old girl by a church minister in Savaii. According to local news media, the minister was allegedly been removed from his post and the victim's family was also banished from the village. Later, a statement from the CCCS Secretary General stated that the church elders have the final saying regarding the removal of said pastor.

In Samoa, a very critical related matter that need immediate attention is the lack of resources and the inefficient legal processes that deal with this issue. Samoa Observer reported during this period that more than 2,000 cases of domestic violence have not been investigated. This damning statistic should trigger government actions to curb the violence and bring closure to the victims and the community as a whole. Insufficientt funding is a real world-wide problem in these types of issues. Take for example, there are thousands of DNA sample relating to criminal cases, that are clumped in hundreds of Law Enforcement freezers all around the US that need to be analyzed.

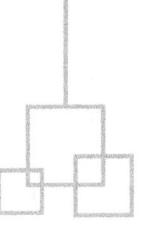

Climate Change

According to the Christian Faith, in Genesis, God created Earth and approved man to be its steward, *tausimea*. Therefore, all Christians should contribute to making sure mother Earth, created by their God, stays healthy. The church has the responsibility to assist governments in making sure future generations can enjoy the magnificent and stunning craftmanship of the Almighty Christian God. To gain a glimpse of nature's beauty, one can enjoy images from the Ark Project and the Coral Morphology company. This natural beauty is referred to in Luke 12:27. Many people use hundreds of dollars watching and admiring nature, e.g., Bird and wolf watchers. One member of our *aumaga* standing in the middle of his large taro and banana plantation, and looking towards his family's beautiful flowers surrounding their derisory *fale* commented that the beauty of nature (Genesis 1:31) was created just for planet Earth and not for other bodies in the universe. My personal belief in the Christian God stems mainly from observing the beauty of nature, not from studying scriptures. Refer to Romans 1:20-22 "For since the creation of the world His invisible attributes are clearly seen being understood by the things that are made even His eternal peace and Godhead...."

One would really be amazed and desire to save Mother Earth after viewing the images mentioned above. Various world-wide programs endeavor to address the negative impacts of climate change with hundreds of people around the world participating in mitigating projects like recycling, coastal cleanups, alternative energy, replanting corals, carbon dioxide recapturing and many more. Thousands more have protested in many countries in support of mitigating Climate Change projects, where many protesters have been hurt and jailed.

It should be noted that man is the main culprit of the negative impacts of Climate Change but nature also seem to have some small contributions to this problem. For example, natural gas emissions have been increasing in the past decades. About half of all greenhouse gas emission come from farms!. Methane from burbing cattle's makes a particularly big contribution[36].

Despite the challenges and pessimism, there was a light at the end of the tunnel when the US Department of Energy announced a scientific breakthrough in nuclear fusion that can result in developing a new, sustainable form of energy that releases virtually no carbon dioxide or other types of air pollution[37]. Global warming became significantly important during this period when Earth reached its hottest day ever 3 days in a row. For the Samoan Islands, this is a serious problem moving forward. The quantity and quality of pacific island climate data has been getting worse in recent decades according to the report Climate Change in the Pacific 2022, by the Pacific Community, and I personally understand the implications and intricacies of this issue as I posses a Certificate on "Modeling Climate Change" from the University of Chicago.

A very interesting conversation has also emerged during discussions of this phenomenon. According to Dr. William Harper of Princeton University: CO_2 has been demonized because it allegedly causes catastrophic global warming, in fact more carbon dioxide will be beneficial and crop yield will increase. This is another classic example of confused atheist-scientists. I'm hoping the science

---

[36]

[37] Lawrence Livermore National Laboratory, California. Dec. 2022

community would come to an agreement regarding this issue but I know they are like the Christian community who are also confused and disagree with each other on religious issues.

In the US, a student asked a professor "How old is mother Earth?". The professor replied: "according to which model?". When people asked scientist Neil deGrasse Tyson if he believe in God. The famous contemporary scientist replied: "which god?". If one asks questions regarding muslims, one could get a reply like: which muslims? those that are progressive?, those that believe in peace or those Jihadists that are suicide bombers?. The same kind of reply can be obtained from asking questions about Christians. Replies vary if one asks about Christians. One might get answers like: are you asking about the Charismatic ones, the conservative ones, the orthodox etc. Maybe the affirmative effects of this practice can bring about some agreements on the topic of Climate Change.

Planned coal and nuclear plant don't seem to curb this issue in the near future. Samoans have participated in related programs in the past two decades and local churches have also contributed to related mitigating local efforts. My previous book also has a substantial section on this topic[38]. According to AP Science Writer Drew Costley (10-5-22), widespread drought that dried up large parts of Europe, the United States and China during the 2022 summer was made 20 times more likely by climate change.

American Samoa had its most serious taste of this phenomenon in the early morning of July 14, 2022 when multiple coastal villages were littered with rocks, corals and various debris from the adjacent coastal waters, littered one the main road. The airport was closed; government offices were closed earlier in the day; some business opened a few hours later than usual as swells overrun barriers along the coastal areas. The government later send food and water for the residents of Aunu'u as a State of Emergency was declared.

On October 4, 2023, a University of Hawaii scientist explained

---

38

on KHJ Radio that American Samoa has experienced **both** sea level rise and the sinking of its islands in the past few years, due to climate change and recent earthquakes. It was interesting since the Cook Islands and Tahiti were experiencing similar weather conditions at the same time. In Samoa, the Ministry of Natural Resources and Environment planted 1,500 trees on the Methodist Reserve in Afiamalu as part of their million-tree campaign. This was a component of a larger effort to promote the mitigation of Climate Change.

Most people think that this topic is relatively new but it's not. In September 1928, climate-change's relation to humanity's future was articulated by geologist Thomas Chrowder Chamberlin, two months before he died. When a journalist interviewed him in his Chicago study, Chamberlin gave a crinkled smile and said he was "a declared believer in large opportunity" for humanity. Chamberlin pointed to the fact that humanity had only just discovered the "enormous energies" corked up in atoms. Our species is like an infant, he continued. "From the standpoint of the Earth, I am an advocate of a great future". In 1899, he proposed that CO_2 causes global warming and also suggested that human activities are altering Earths future climate. In June 1898, a onth **before** Curie introduced the term *radio-activité* – Chamberlin was asserting that our ignorance regarding subatomic processes means that we should be suspicious of Kelvin's estimates of a meagre future"!.

A relatively new approach to measuring global warming has been suggested by scientists from the University of California San Diego's Scripps Institution of Oceanography and Cornell University. They suggested that humidity matter in generating dangerous climate extremes, not just temperature. By adding the energy from humidity "the extremes – heat waves, rainfall and other measures of extremes – correlate much better", according to these scientists. In a related recent UN report: *Climate Change 2022: Impacts, Adaptation and Vulnerability,* the authors stated that "many of the impacts of global warming are now simply "irreversible" but there is still a brief window of time to avoid the very worst, and that 40 percent of the world's population are "highly vulnerable". This include the Samoan Islands where some of the coastal areas of Aunu'u Island and villages near the airport in *Faleolo,* have already been impacted. This report is the second of three reviews from the world's foremost body of climate researchers.

The southern winter that just ended during this period in New Zealand was the warmest ever recorded, and scientists say that climate change is driving temperatures ever higher (AP, Sept 9, 2021). For the three months through August, the average temperature was 50 Fahrenheit, according to New Zealand's National Institute of Water and Atmospheric Research. That's 1.3°C above the long-term average and 0.2°C higher than the previous record posted last year. Scientists have been keeping records since 1909, but most of the warmest winters have been recent. According to the Institute, on top of a background of global warming that year there were more warm winds than usual from the north and warmer sea temperatures and that the underlying warming trend can be tracked through carbon dioxide concentration, which has increased in New Zealand from 320 parts per million 50 years ago to about 412 parts per million today.

After 15 years of research the Marine Mammal Center and scientists in Australia have discovered that the devastating disease in coastal dolphins worldwide is caused by Climate Change (Katie Camero Dec. 22, 2020). Frequent storms and severe global temperature rise are to blame. Large volumes of rain from storms are poured over saltwater oceans, slowly turning them into freshwater reservoirs. The decreased salinity causes patchy, raised lesions (condition called ulcerative dermatitis) over the bodies of dolphins (Scientific Reports: 2020).On the lighter side, Victoria University pointed out that farmers with cow or sheep herds might benefit from a longer grass-growing season.

The public should be aware of "greenwashing" which is a relatively new phenomenon where businesses claim that their products are environmentally friendly or organic. Some of these claims maybe true and some are misleading. This has confused many consumers as they have no way of verifying these claims.

Environmentalist in Scotland recently sounded the alarm in hopes of preserving one of nature's best weapons in fighting climate change: peat bogs. Peat bogs make up just 3% of the Earth's land mass but store 30% of all land-based carbon, which is twice as much as all the world's forests (Mark Philips. 2022).

According to the Associated Press (AP 4/22/22) scientists say last summer was the hottest summer on record in Europe, with temperatures a full 1 degree Celsius higher than the average for the previous three decades.

For those readers who are interested in local related events and observations, I had provided some **local** events and related perspectives on this subject in my previous book. Some events that might be of interest to the readers (maybe not directly associated with Climate Change):

- In July 2022, the Samoa Meteorological Service announced that "three earthquakes were felt in July with two occurring on the same day, but the tremors we hear about are just a tiny fraction of the actual amount recorded each month ".
- Occasional downpour while the sun is shining brightly as the one, I personally observed in Utulei on August 11, 2022 around noon.
- In August 2022, the thermometer dipped 70 degree which was a record-breaking low temperature for American Samoa.
- *Major damages in Apia harbor costing millions of Talā to the Friendship Park, in early 2024,* were caused by extreme tidal surges and ocean currents, a very rare event. A few ships, including an Ocean Liner and the vessel Manu'atele from American Samoa weren't able to use the main Apia wharf, and a fishing boat also ran aground in that area. At the same

time, large waves brought in lots of ocean litter onto the Faleasao harbor, Manua.

Similar events also happened around the same time in the Marshall Islands which cause major damages to the Freeflight International Airport and the Dyess Army Field.

The American Samoa Government in September 2022, held a special church service at the Iakina Seventh Day Adventist Church in Iliili to kick off the Coastweeks Celebrations under the theme "A Healthy Coast is a Wealthy Coast". The Governor reminded the audience that "this celebration is a reminder to us all to be proactive stewards in our efforts in protecting and preserving what God has gifted us". He also added that his administration will help manage, sustain and protect the coastal zone and marine environment so that future generations of American Samoa will also benefit and prosper. The service commemorates the 50th. Anniversary of the American Samoa Coastal Zone Management Act. This is another example of how the government and the public view the environment as a God-given gift that must be protected.

Considering the scientific phrase that for every action there's an equal and opposite reaction. Now ponder this. Humans have been polluting mother Earth for decades, that's the **action**. Hurricanes, droughts, storms, earthquakes, volcanic destruction, and all the unforgiving climate conditions are the consequence **reactions.**

This problem should be viewed and "attacked" in a **holistic** approach to be successful. Much like the medical holistic approach, the emerged in recent decades, where various approaches like yoga and prayers are used. Engineering, (novel approaches utilizing better systems), ecological considerations (e.g., utilize marshes and mangroves more); commerce (consider constant waves from huge ships battering and changing the coastlines). Countries should also look and learn from countries like Venice and then formulate their own approach.

The environment is also very important in politics. Some

researchers believe that one of the main factors that contributed to Sadam Hussein downfall was his poor decision to drain the marshes. This event caused conflicts with Syria. With the current low level of water of the Euphrates, it should be noted that Jeremiah and the Book of Revelation prophesied this condition. In the US, GOP members including think that Climate Change is a hoax. Lets see if that idea hold up in the upcoming Presidential election.

There are several natural cycles in nature. For example, the oxygen, nitrogen, water, citric acid cycles and many more. The are many people who believe in the cyclic nature of climate and this is true, but human activities have contributed significantly to the **frequency** and the **intensity** of climate change effects in the past decades. **These are the foundational ideas of the modern concept of climate change - frequency and intensity.** Debates against Climate Change should however keep certain questionable characters out of national television. For example, Professor Willam Harper stated on the Institute of Public Affairs that solar farms should be called "solar plantations" and that these don't produce electricity at night and have no economic reason". A very weird observation.

There are also very several prominent members of the public that have campaigned for many years against climate change and consequently provided false information that had "muddied the public's understanding of climate change". According to "World. The Cool D. TCD. Collins, Leo", Robert Murdoch, assisted by columnist Andrew Bolt, has wielded his global media empire as a cudgel to sow confusion and doubt about the science and the solutions". May I add, and for this "sin" against God's planet Earth, he had to resign as head of his organization.

For the non-scientific community, an every-day explanation of this topic can be found in "The Physics of Climate Change", an online lecture with Lawrance K. Krauss which is part of the Origins Project Foundation.

Recent data from the Pew Research Center indicate that the

majority (54%) of American adults describe climate change as a major threat. The data also show a growing partisan divide: "while 78% of people who leans towards the Democrats believe it's a threat, only 23% of those who lean Republican feel the same way". Furthermore, there is evidence that people have been changing their minds. In 2018, research from George Mason and Yale universities found that 8% of Americans had recently changed their opinions about global warming – the great majority of them had become more concerned about it.

The various denominational implementation of baptism should ponder the decades-old predicament of worldwide droughts which has increased in recent years due to climate change. Lakes in the US and other countries like India have experienced this serious problem and have caused environmental and health damages to millions of people around the world. A recently reported example is from the Khadimal village in western India where "people risk their lives every day for one bucket of water'. If I were a resident of that village, I wouldn't join a church (e.g., AOG) in that village that practice full immersion. I would use the meagre amount of available water for my family to drink, prepare food and bath. In this situation, I don't think full immersion is God's will. Additionally, it's just a symbolic indication of a more spiritual concept. This water problem is not a problem for the past, it is actually happening **right now** in communities like Monument Valley Utah and in Arizona and many other parts of the world.

In these islands, many local churches, schools, NGOs and the government are willing to help programs to fight this problem. The issue however, is similar to the concept of Christianity, in that it is and will always be challenged as analysis of new and historical data are analyzed. Prominent figures like Al Gore (Former US President candidate) seem to suggest that humans should make several sacrifices to ensure the survival of life on Earth. Others disagree and noted that humans don't need to sacrifice the comfort of their lifestyles but should improve technology to combat Climate Change.

Some scientists suggested that the climate in the past several decades is not the norm but the exception. There are scientific data for and against all these arguments and the discussion will continue onto the next decades since prolonged and comprehensive historical and current data is required to make sound determinations for this issue. The Christian God, however, had created a "very good"(according to Genesis 1:31) place for mankind and I truly believe humans will continue to enjoy this planet for a very long time if they treat Mother Earth with respect. Pray for her my dear friends.

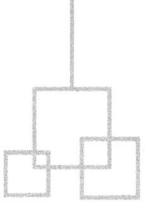

Not for Profit Organizations and Good Samaritans

T here were several Not-for Profit Organizations (or Non-Government Organizations - NGOs) operating in the Samoan Islands during this period (an extended list was included in my previous book). These organizations provided services and various assistances similar to those provided by the various churches, and therefore their inclusion in this book. Governments of both Samoa and American Samoa also provided various related assistances for the community on several occasions in the past three decades.

After a year focusing on leadership training, American Samoa Mission (ASM) Pathfinders, Adventurers, and Youth had the opportunity to put their knowledge into practice, leading and participating in a community service project that helped a variety of organizations in American Samoa. Conducted on a Sabbath afternoon, the project distributed food, bedding, and toiletries to government facilities, such as the Tafuna Correctional Facility, the Juvenile Detention Center, Hope House (home for the elderly and disabled), and two shelters housed under the Department of Human and Social Services (DHSS). Representatives from the different organizations expressed their gratitude to the youth groups that presented them with donations.

Through the Civil Society Support Programme, the Australian

Government provided SAT$173, 000 to SENESE's Inclusive Education Support Services in October 2021. The funding was provided to strengthen support services under Hearing Services, Speech & Language Therapy and Vocational Life Long Learning. The funding was actually awarded by CSSP in August 2019. This type of funding was to assist NGOs address the needs of vulnerable people including those with disability. SENESE is an NGO that provides learning and support for children and youth with disabilities in schools and in the communities.

Patients at the Tupua Tamasese Meaole National Hospital Pediatric Ward, received various gifts from the Faith Angels NGO which included hygiene kits, food, nappy wipes and other items. According to the organization their service is based on: "Faith is the substance of things hoped for, the evidence of things not seen (Hebrews 11:1)". The foundation was formed following the 2019 measles outbreak.

Hundreds of families with children in California and Pennsylvania were surprised with gifts and more when the Pittsburgh Steelers' Samoan wide receiver JuJu Smith-Schuster paid off their expenses on lay-away items for Christmas.

After the volcanic eruption in Tonga in January 15, 2022, several overseas NGOs, companies, governments and private citizens donated to help the people of Tonga. Many local people, the two governments and churches also donated money, food and various items. The American Samoa government provided nearly half a million dollars plus other items. An interesting assistance came from the very rich Elon Musk, who installed 50 VSAT terminals in the islands to help the disrupted communications systems.

During the Code Red restrictions of the 2021 Pandemic, four NGOs, Rotoract of Pago Pago, Esther Generation, Agency for Better Living and the Christopher James Foeoletini Foundation joined forces in assisting low-income families in American Samoa with their electricity bills via cash power payments. According to this initiative organizers, KS Inc. donated $2,000, CostULess and

Marvin Rapps donated $2,000, Bluesky Communications donated $2,000, Tropical Beverages donated $1,500, Tropical Pizza and Tropical Chicken donated $1,500.

The Samoa Victim Support Group (SVSG) and the European Union collaborated under the Civil Society Programme II in support of the "Nofotane Women's Seminar" on March 3 and 4th, 2022 at Savaii. The theme of the program was to foster partnership and develop networks between village representatives and the self-employed *"nofotane"* women in creating a meaningful change in sustaining their income generating activities.

The Pacific Leadership Forum(PLF), an assembly of Pacific organizations and leaders provided RATS (Rapid Antigen Test) to several communities and especially churches in Aotearoa to help with the COVID-19 pandemic. This accentuates the participation of the churches in community issues. PLF was also monitoring the situation in Samoa in anticipation of future assistance.

In 2022, the Samoa government provide $US279,940.0 to the Samoa Victim's Support Group (SVSG), the largest grant for an NGO in Samoa. The funds were to be used for the elimination of violence among Samoan families from its headquarter, Campus of Hope. According to government statistics at this time, it was estimated that 46 percent of women experienced some kind of abuse from their partners.

A group of children housed at the Samoa Victim Support Group (SVSG) Campus of Hope House, Tuanaimato celebrated 2022 Easter Sunday with special praise to the love of God that gave them a second chance in life. The theme of their celebration was "Praise and Honour to Jehovar". The children range from babies to 16-year-old who have gone through either abuse in the hands of family and close relatives or abandoned by parents.

The Lions Club of American Samoa donated three truckloads of humanitarian aid to the Territorial Correctional Facility in Tafuna in April 2022. The items donated included food items, clothing and hygiene items.

The Australian Salesian Mission Overseas Aid Fund donated ten industrial sewing machines valued at $15,500 SAT to Don Bosco Technical Center in May 2022. Lauren Bicknell Hichaaba and Tina Newton were instrumental in organizing this donation. According to these two organizers, "the sustainable focus of the program enables support for the school so they can continue to provide top quality education for young people in Samoa". The donation was to "upskill the students and also the community at large".

The Nofotane Project donated new computers valued at $4,500 for the Children of SVSG and was through a request for computers to enhance the learning and education of the children living at the Campus of Hope.

The new FAST government allocated $500,000 specifically for active non-government organizations that offered health services. These funds were for programs to combat communicable diseases, vaccinations, cancer, COVID-19 response efforts and capacity building for the Health Care sector. Related NGOs included Samoa Cancer Society, Samoa Red Cross, METI, GOSHEN Trust, Samoa Family Health Association and AGAPE Clinic.

In June 2022, NGO *Samoa La'ala'a* gifted SAT $10,000 to the *Mapu I Pulotu Home* for the Elderly in Savaii. This NGO was formed in 2000 and the group hope their donation would help ensure aged care services are provided for the Savaii elderlies without travelling to the *Mapuifagalele* in Upolu.

In 2022, the Aiga Malamalama Tatau Club donated $1,500 tala to Mapuifagalele. The new NGO lead by tattooist S. Paulo Suluape Lafaele has more than 40 members, all wearing the traditional Samoan *laei*. The organization was started by former students of Avele College and was formerly known as the *Avele Malofie 2009*.

In August 2022, the SSAB store donated 400 Bibles to the Tanumalala prison. Officials attending the event included Senior Judge Vui Clarence Nelson and the British Charge d'affaires to Samoa. Owner of SSAB's mother, Manino Stanley Fuimaono, had been donating to the prisons for many years as an individual and as member of Fortress of Faith NGO.

The CCCS Youth Group of Siutu Palauli and members of the church were gifted with 21 solar lights in 2022 from the Samoa Victim Support Group. The lights would light up the playing field of the youths so that the youths would be able to utilize the playgrounds at night, not only for their young church members but for other youth organizations in nearby villages who need playgrounds at night. The solar lights were valued around $5000.00. Since the energy will be from the sun; maybe the angels should appear every month to collect the associated electric bill only from atheists youths who use these playgrounds?. Ask your pastor.

The Rotary Club of Pago Pago provided furnishings to the Hope House and the Hope House's new Adult Day Care Service in September 2022. Various items, bought from local businesses were presented by ten of its members. The donated items included 18 wall-mounted fans, a large refrigerator and 3 recliners.

In 2022, The CCCS Women's Charity Fellowship, *Tautua mo Tagata Puapuagatia (TPT)*, presented $70,000 to the Minister of Health for various health-related projects. $10K was for the National Hospital, $10K for hospitals in Savai'i, $30K for the National Kidney Foundation Moto'otua and $20K for the National Kidney Foundation, Tuasivi, Savai'i.

The Resident Representative of UNDP and staff visited the children at the SVSG Children of Hope around Christmas 2022 and provided various entertainment.

Neil's ACE Home Center in American Samoa has been partnering with the American Samoa Cancer Coalition to raise stipends for cancer patients for their medical expenses, medical care and daily expenses since 2009. In 2020 it donated $33,000.00 to the Cancer Coalition.

The LDS Charities, the humanitarian arm of the LDS church, gifted a 7-year-old from Sinamoga in April 2023 with a much-needed wheelchair. This resulted in a great positive change to

Sandra's everyday life and was much appreciated by the family, but now there's a "problem", Sandra's siblings keep bickering about who gets to push the wheelchair. This is the type of work atheists should be involved in instead of wasting their energy in debating the Christians.

The Samoa Victim Support Group (SVSG) in early 2023, received several donations from the young students and teachers of the Vaiola Primary School of the Church of Jesus Christ of Latter-Day Saints. The donation was specifically for the children housed at the SVSG Campus of Hope, a shelter for abused children. Obtaining the donations was launched through the school's "Mini Mission" initiative, where various items like toiletries, clothing, detergents and food items were donated by the students and teachers. The donation was received by the SVSG branch at Tuasivi, Savaii.

A project called the "Agabe Homes Project" aimed at building better homes for the poor was in its second stage during this period. Its pastor Rev. Nofoaiga Eletise, and members of his church, South Seas Christian Ministries church based in Australia, were in Apia in April 2023 to continue arrangements for their project. The project aimed at starting to build homes in 2024.

In 2023, the Electoral Constituency of Gagaemauga #2 gifted its Church Leaders with new printers, laptops and printer papers utilizing funds from a FAST government project. The items cost around T$33k.

God's Power Ministry based in Sydney Australia under the leadership of Pastor Sam Tufuga, during its Samoa Mission Conference 2023, executed a medical screening clinic at the Tupua Tamasese Efi Building and presented the Ministry of Health with boxes of medical equipment.

In July 2023, a group of doctors and specialists from the US working with the local Lions Club of American Samoa conducted free clinics on Tutuila. The public was welcomed to see eye doctors and specialists for a few days for free.

DMC Restaurant and Coin Save donated a 3-piece children's

playground equipment and accessories to the SVSG Campus of Hope during this period. The playground caters for children aged 2-5 years. This will include abused, vulnerable and those with special needs.

The Faasao High School Alumni Association visited the Hope House in Tafuna in July 2022 and spend quality time with elderly and disabled residents. The Alumni engaged the residents with activities like Bingo and *Elei* painting. The Association also provided towels, bed linens, laundry supplies, cleaning supplies, bed linens, snacks and a monetary donation.

The New Zealand-based Sanitarium Health Food Company, with assistance from the Adventist Development and Relief Agency(ADRA) hosted a breakfast treat for 25 Primary Schools from around the country in July 2023. Sanitarium is the leading health food company in New Zealand, the producer of the popular Wheat-Bix products. ADRA is the global humanitarian organization of the Seventh-day Adventist Church, founded in 1956 as the Adventist Welfare Service and renamed ADRA in 1984. The organization also donated $10,000 worth of linen to the National Hospital at Motootua and many gift bags for the children.

American Samoa's LBJ Hospital received a donation of bedding materials from two stakes of the Church of Jesus Christ of the Latter-Day Saints in July 2023. The donation was part of the stakes' community service and included blankets, pillow cases and spreads. The donation was from the "Sisters of the Relief Society" of Pago West and Pago Mapusaga stakes. According to church parishioners from these stakes, there are five stakes in American Samoa with more than forty Wards and branches. The donation was a timely and well appreciated by LBJ since there is always a shortage of hospital bedding supplies according to the Human Resources Director.

The missionary group Island Breeze visited American Samoa during this period for its outreach mission. The group was formed in Nuuuli village forty years ago to preach the gospel through music and dance. Great Life Church 684 hosted the group who was to visit schools and had planned two public events during their visit.

Samoa Ports Authority continued its Annual Charity work in September 2023 by donating monetary gift of $5,000 and a variety of supplies to the *Mapuifagalele* home for the elderlies, which were received with much appreciation by the Sisters of the Poor.

In 2023, the US Navy ship, USS Jackson delivered aid and provide various development assistance to communities across Samoa. The vessel had a multinational force of engineers, medical specialists, musicians and sailors. This was part of an annual humanitarian operation known as the Pacific Partnership.

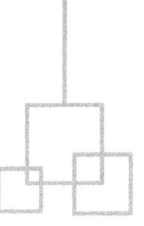

LBGTQ+

P ersonally, this is the subject that I utilize to separate the Christian "boys" from the Christian "men". For example, I used to watch Pastor Gino Jennings on television but then when he preached against the LBGTQ community, I now refuse to watch this preacher. Since he said he never attended any Bible College, I will approach Malua and Piula Bible Colleges if they would kindly accept this preacher to study there.

The Bible has teachings regarding this subject and the Christian community world-wide have varying assessments and perspectives on this subject. My general position on this issue is a personal question: If *Agabe* and God's all-inclusive love doesn't include this group then I'm baffled and lost towards Christianity. Moreover, if these people have absolutely no saying in what they would become as humans – absolutely no fault of their own – and the Christian teachings is against them, then this issue has diminished my respect for Christianity. But I believe God embraces all kinds of people. Consider Timothy, who was described as "half-bred (his father was Greek) and "should have never been born" was allowed by God to serve him. He also wrote two Books of the Bible. I encourage the readers to talk to members of this group (not to the ill-informed pastors) to obtain the essence of this matter. I encourage pastors who preach against this group, to talk with these people, for example the Rogers in Samoa, and learn directly from them and not read text books or wrongly interpret the Bible. Who are you to judge silly, said one chief of my village.

Samoa and American Samoa both have notable LBGTQ organizations that have contributed to charities and the community for many years and several members of these organizations are respected members of the community. The Samoan church community generally preaches varying ideas regarding this group. A late Superintendent[39] of the AOG church often preached against this group while government(s) and the public seem to tolerate and accept this category of humans. Rev. Asiata of the Matuu and Faganeanea CCCAS, pointed out during one of his church's Christmas programs that members of this group have been unfairly treated by the community for decades and it is now time to "correct this wrong". According to a Catholic Deacon form Pavaiai, he personally doesn't look down nor condemn members of this group.

According to recent research (Tony Morrison. Feb. 17. 22) in the US, the number of adults who identify as LBGT has doubled in the past decade. Some Bible scholars noted that Jesus never condemned nor mentioned homosexuality, some disagree. A new film "1946: The Mistranslation That Shifted Culture"[40] alleges that the Biblical assertion that "homosexuality is a sin" was a **translation mistake**, and has resulted in the "stoking of homophobia – a mindset of hatred, oppression and religious nationalism that has defined the last 75 years". For the two words that were examined by the film's research, one accurately means effeminate and the other connotes a person who was a sexual abuser and who had harmed someone. An attempt by scholars to correct this issue was hindered by the AIDS crisis in the '80s.

In the Samoan islands, during the past two decades, there was an increase in equality, representation and visibility of this group. Overseas churches like the Charlotte First Methodist Church, which

39
40

officiated its first marriage of two men during this period, have increased the acceptance of this group.

A few instances that overseas reporters were able to document showed that being gay may help save one's life.(I personally feel that there are hundreds more similar stories that aren't being reported). Take the story of a Denver male couple who found out that one of the partners was an ideal kidney donor; a surgery was performed and saved the partner's life. Alport syndrome, the related culprit, affects 30k to 60K in the US. (OUT NEWS Dan Avery 9/23/2021). President Joe Biden's Secretary of Transportation and also former Mayor of South Bend, Indiana would have been the youngest and first openly gay President of the US, according to an Amazon Studios documentary about Pete Buttigieg, which was scheduled to be shown during NewFest's 33rd. edition of the New York LGBTQ film festival. (Variety. Sept. 2021). He was later appointed by President Biden to a cabinet post. Sturt Delery was to be the first openly LGBTQ person to serve as White House Counsel in 2022.

During this period, the Evangelical Lutheran Church in America installed its first openly transgender bishop in San Francisco. (Washington Examiner 9/15/21) and several talented members of this group received awards during the September 2021 Tony Awards event in the U.S. Also, during this period, The Church of England's National Assembly voted to let priests bless same-sex marriages and civil partnerships, while continuing to ban church weddings for same couples.

Right Rev. Gene Robinson in 2003 became the U.S. Episcopal Church's first openly gay bishop. Bishop Tutu had supported him. Tutu, South Africa's Nobel Peace Prize-winning activist for racial justice, died at age 90. Bishop Tutu was an uncompromising foe of apartheid, South Africa's brutal regime of oppression against its Black majority, as well as a leading advocate for LGBTQ rights and same-sex marriage. In the foreword to Robinson's book, Tutu also apologized for the "cruelty and injustice" the LGBTQ community had suffered at the hands of fellow Anglicans.

Chile had its first men couple legally married in March 2022.

One of the problems causing unnecessary stress and conflicts regarding same sex marriage and related problems is the mediocre standpoint made by some members of the clergy, prominent members of the community and petite officials making related decisions that should be done by senior representatives of associated entities. For example, a US judge ruled that a former county **clerk** (just a measly clerk) from Kentucky knowingly violated the rights of same-sex couples by denying them their Marriage License in 2015. The clerk cited her **personal** religious beliefs (Reuters 3/19/22) but was later jailed. In 2023; a judge ruled that she must pay $260,000. Under the US Constitution, states must grant same-sex marriages and recognize those of other states. Supervisors, judges, attorneys and other higher officials should make related decisions, **not diminutive clerks with no degrees.**

The 2022 59th. Venice Biennale, one of the most prestigious art and cultural exhibitions in the world showcased a series of photographs by Samoan-Japanese *fa'afafine* artist Yuki Kihara which explored the idea of *fa'afafine* utopia. The photographs "upcycled" selected paintings by Paul Gaugin. Samoa Fa'afafine Association mentioned the idea of pre-Christian portraits and paintings of native Samoan women in regards to Christian values and perceptions introduced by the LMS in 1830.

The Pacific Sexual and Gender Diversity Network urged pacific island governments to stop discriminating against those in the LGBTQI (lesbian, gay, bisexual, trans, queer, intersex) community during its International Day against homophobia, transphobia and biphobia in 2022.

A critical concept that all Samoan pastors need to grasp is that members of the LBGT group had nothing to do with the way they were born into. The US National Institute of Health (NIH) had acknowledged the existence of natural porn or sex impulses in some humans. Should we then rebuke these people all because of a medical condition?. An LBGT friend of mine told me that at night, he/she would lie in bed and while looking up the ceiling would ask God "Why me? ". Imagine going to prison on a charge you never committed. Imagine being born with no limbs, one eye, deaf or as a conjoined twin (all these are true). Was it your fault? Several US government programs that tried to "correct the behaviors of these people have **all failed** because the root cause is in the DNA. So is God or evolution to be blamed?. Absolutely not (come by to discuss this with this author, a self-proclaim expert in this field). Moreover, the Pope said "homosexuality is not a crime". The Pope also said that "God loves all his children just as they are" and he also welcomed LGBTQ people into the church. **Samoans should stay away from churches that preach against this group of people that God created in his image**. The Bible also teaches "not to judge others". Some local pastors preached that it's the behavior of these people that is immoral. Do they expect a horse to behave like a snake? Both biological creations of the Almighty?. Please grow up and expand your horizons and perspectives. This is the 21st. century and man landed on the moon in 1969 for God sakes. A previous Superintendent of AOG once preached against the LBGTQ group saying that the related wicked behaviors of this group originated from Sodom and Gomoro and that all these sinful people will surely go to hell. A few months before he passed away, he changed his general Biblical views during a service I attended.

In 2023, the President of the National Council of Churches, Aisoli Iuli was reported saying that "if Samoa accepts the advice of Pope Francis to have legal recognition of the LGBTQ community, then same-sex marriage will follow dooming Samoa in the eyes of God". Well Sir if you mean dooming the Samoan Islands will be like what happened to the Egyptians mentioned in the Old Testament, then I humbly request that you invite me to your bedside before you pass away. I need to whisper to your ear that ":you were completely wrong".

I have also made a similar approach to the leader of the Independent Seventh Day Adventist Church, that I will visit him in about 5 years and whisper to his ear that "you are still alive, I told you so", after he made public comments that his blood will flow in the streets of Apia because people will assassinate him for telling the Biblical truth. His remarks infuriated the Prime Minister and many denominations. As a very confident writer; I will keep my promise. People, like the prophets and missionaries like John Williams who brought Christianity to these islands are sometimes killed for telling the truth. People who tell the opposite are mostly not killed!. As I wrote in a few sections of this book, any reader who wishes to debate me on any issue in this book, or in my previous two books, are welcomed to visit me and debate those issues. I had debated one of the top atheists (via email) in the world (Dan Barker) and ready to dabate anyone on this issue.

My other comment on this issue, as a scientist, is that feelings, yearnings and predicaments of these people are caused by normal human conscious and natural bio-chemical processes. Simply put, it's like getting thirsty or hungry. One simply has to obey the natural "request" and this is exactly why several shocking cure therapy programs have failed to curb these natural urges. Therapies like Shock Therapy, Castrations, Torture drugs, Hypnosis, Conversion therapy and even Spiritual Therapy and others have **ALL failed miserably**. During this period, Michigan lawmakers gave the final approval to legislation banning so-called conversion therapy for minors which has been deemed as scientifically-discredited. If one's body wants water, one can't provide any therapy to cure this, one just has to provide the body with water – what it wants. Its natural, not religious. Let me explain this further for the 'young" readers, from a legal perspective. These people didn't have any saying in what they will grow up to be and what is in their DNA. Therefore, I can liken them to those who have been accused of a crime they didn't commit! To those pastors who are on their high horses and need help in changing their views on this subject, just talk to ex-convicts who have been incarcerated for 20 or more years for crimes they didn't commit. I can help you locate scores of these people in the US or just Google the Innocence Project.

In Samoan contemporary literary work, a Samoan writer paid tribute to her *faafafige* friends in her play *O Tusitala, Teller of Tales. a* play about the famous Scottish writer Robert Louis Stevenson who lived his last days in Samoa and is buried on *Mount Vaea*, on the island of Upolu. On a related note, Robert Louis Stevenson's short story "The Bottle Imp" was first published in Samoan as "*O le Fagu Aitu*" in the London Missionary Society LMS – now EFKS publication "*O le Sulu Samoa*" before it was published in English, in 1891.

For those, especially members of the clergy, who have contrasting views, let me provide an additional perspective and analysis. A new Documentary called "*The Mistranslation that shifted a culture*" has shocked many modern-day Christian fundamentalist and those who have no tolerance for homosexuality. According to this documentary, the Bible never mentioned homosexuality until 1946. Poor grasp of some obscure Greek terms and ignorance (may I also add "arrogance") later added this notion. According to forgotten Yale University archives, the mistranslation was earnestly spread by Conservative Christians in the 1970's in order to scapegoat gays and combat their burgeoning liberation movement. According to this argument, none of the older Bible versions contain the word "homosexual" and that the Bible actually condemns people "who are lazy, weak, slothful and predatory", not homosexual. The documentary noted that, to the 1946 translators, that was meant "homosexuality", which according to this documentary was "their spin, NOT GOD's. Readers may also be interested in a book with a similar tone: *The Unheard*, a Deaf Women Regains Hearing in the Deathly Silence of Cape Cod.

In 2023, controversies abound regarding this group getting the traditional *pe'a* and/or *malu*. One Master tattooist, Suluape, believed that this group shouldn't have these tattoos. Some disagreed and questioned who decides these types of cultural features. Some people noted that there are many non-Samoans who have these tattoos and that there are also several non-Samoan tattoo artists in other countries. Like any subject, there are various opinions out there and for me, I'll leave this to "personal judgement and preference". As my daughter is planning to have a *malu*, I have obtained some old pictures of traditional *malu* designs that the Germans[41] documented when they were the Administrators of these islands in the past. This issue was also highlighted in American Samoa when a team of female Community College students performed the traditional welcoming Ava ceremony for a delegation from the US who visited these islands for a Mental health Summit. This formality is normally performed by males. So who decides the rules of a culture. Come by my house for some Koko Samoa and Ill explain.

I personally understand this topic well as I have friends and relatives who are members of this fine group of people. I also **pray earnestly** that Samoan pastors who shunned these people, may in the future have lots of gay and lesbian children and grandchildren so that they could **personally** understand this issue, for who am I – a lowly, meek janitor and custodian by profession to lecture these fine intelligent clergymen. In case readers would wonder if I'm the only idiot sticking out his head for this group, well, professor Jeremy Atherton, the world's foremost expert on the topic of homosexual, Professor Bart Eherman, the world's top researcher on the New Testament and renown scientist Dr. Neil deGrasse Tyson seem to share my view on this subject. From our shores are the Pastor of the *Matuu and Faganeanea* CCCAS, a former lecturer at the Kanana Fou Bible College (Mafo'e) and a Catholic priest from the Western side of Tutuila, to name a few, agree with my perspective. Furthermore, I think Jesus never mentioned homosexual. Amen.

---

[41] Samoa Islands, Kramer Vol II. Polynesian Press.

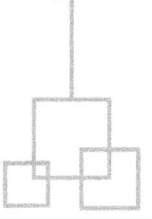

COVID-19 Pandemic

T he COVID-19 pandemic of 2021 caused serious social, economic and medical problems to the island community. Samoa and American Samoa both declared lockdowns for varying periods. During the pandemic, Samoan Churches adapted by utilizing new technology like Zoom to offer online services. Some churches, like that of the AOG Antioch church, under the leadership of Rev. Asaua Fuimaono, sent out assistant pastors, lay preachers and deacons to local families to conduct holy communions and pray with families within the villages.

Very limited number of parishioners were allowed in to some of the churches and social distancing was observed by some denominations. Church services were generally disallowed during lockdowns. Yet, some unconcerned pastors disobeyed the restrictions and police were forced to stop the services in a few churches. One church in Samoa was fined $5,000.00 and I predict that God will not pay this fine since He didn't condone this stupidity. If in the future God personally pays this fine, then someone please lookup my contact information in this book to contact me so I can personally pay the fine instead of God. I swear. As one of my pastor friend (Rev. Filipo Tuigamala of Fagasa AOG) said "God is not stupid".

During the 2022 Easter weekend, several churches in Samoa started their Sunday services earlier than their normal schedule to ensure they finish early prior to 12 PM, to abide by the government's

COVID-19 restrictions. The country was under Alert Level 2 and church services were limited to 30 people. This included the Apia Protestant Church and the New Wine Worship Christian Center Church. The ministry of Police also hosted an Easter Sunday service at their headquarters. Parishioners were asked to provide a COVID-19 vaccination card, observe social distancing and wear masks for the entire duration of the services. At this time the cumulative total of those infected was 5, 074 and 10 deaths were documented by Samoa's Ministry of Health.

American Samoa later got its first related casualty in March 23, 2022 where a 44-year-old man with a few underlying conditions passed away at LBJ. As numerous local First Responders work tirelessly for countless hours, some of them reflected that "the crybabies anti-mask people should welcome the very easy sacrifice of wearing masks, but unfortunately, have erroneously compare these mandates to climbing Mount Everest", and it really frustrated them. The pandemic affected people of all walks of life. In Nauru, a 40-year-old Commonwealth Games gold medalist and Olympian died due to COVID-19 related complications.

In the Pacific, ten Fijian church ministers of the Christian Mission Fellowship Church resigned because they did not want to get vaccinated. The ministers were not forced to resign according to their media director. Other churches including the New Methodist Christian Fellowship have had all their ministers fully vaccinated (RNZ 9/2021). The main push for **herd** vaccinations in any community, including the Samoan Islands, is to protect the **whole** population and when church minsters refuse to get vaccinated, they are apparently saying they **don't care about other people** and I think these **selfish** people should not be church ministers. According to Obery M. Hendricks Jr. a religion professor at Columbia University (R Quay. Weekend Editor. 2.14.22) "Christians who refuse to wear masks are possessed by the spirit of antichrist". The Samoan Islands were blessed during the COVID-19 pandemic since it didn't have any pastors with the spirit of antichrist, or were there?. At least none was reported or they were distancing themselves from their main denomination's beliefs. According to the professor "the term antichrist may also mean "forces that opposes Christ and his teachings". This is why I believe these pastors who resigned shouldn't be pastors because they oppose Christ and his teachings. For the local Christian community, don't worry about these pastors leaving, there a scores of more qualified and informed graduates of Malua Theological College, Kanana Fou and Piula Theological College available. This is like the Manu Samoa rugby team training for the World Cup and has several top players as reserves.

During a 2021 trip by the Pope, he said on the plane that he "didn't understand why people refuse to take COVID-19 vaccines" adding that "serious discussions about the shots was necessary to help them". He also pointed out that a cardinal who didn't believe in vaccines was hospitalize and referred to him as the "poor guy". Additionally, he mentioned how humanity has a history of friendship with vaccines, pointing to vaccines against polio, mumps and measles. The Pope didn't cite the ambiguous religious objection used by some confused christians (note my small letter "c"). Some conservatives had used the indirect connection to lines of cells derived from aborted fetuses to refuse vaccinations against Covid19, but the Vatican's doctrine office has said it is "morally acceptable" for Catholics to receive COVID-19 vaccines based on research that used cells derived from aborted fetuses. Pope Francis also said it would be "suicide" not to get the jab and both Francis and Emeritus Pope Benedict XVI have been fully vaccinated with Pfizer-BioNTech shots. I hope all the Catholics all over the world, including those in the Samoan islands, had obeyed their wise leaders. This is one of the issues that prompt atheists to make fun of Christians and portrays the absence of altruism within the churches. Being inconsiderate of others' wellbeing is truly un-Christian. What a shame. Its good these "kids" resigned. Around this period, the largest group of atheists in the US called the Freedom from Religion began presenting commercials on national televisions and was broadcasted in the

American Samoa KVZK TV, maybe they were sending a related message.

Several overseas pastors repeatedly called COVID-19 a hoax. This included Greg Locke, the pastor at a Nashville-area church (Peise, Washington Post. July 27, 2021) who also undermined emergency mandates and refused to comply with guidance from Public Health officials. During a sermon a month earlier, Locke called President Biden a fraud and "a sex trafficking, demon-possessed mongrel," a reference to QAnon, an extremist ideology based on false claims. He also falsely claimed that vaccinations were made of "aborted fetal tissue.". Many overseas members of the clergy had similar perspectives and they **later died** of COVID-19 complications. To me this is suicide and it's against the scriptures. A parish priest Don Paolo Romeo of the Santo Stefano Abbey in Genoa died of COVID-19 after claiming that the COVID-19 vaccines were made using cells from aborted embryos. This theory was debunked by the US Conference of Catholic Bishops, and the Vatican even stressed that Covid-19 vaccines are "morally acceptable". Several US pastors not only condemned the government's initiatives to vaccinate the public but they also stole funds from COVID-19 programs, like Pastor Evan Edwards and his son who were arrested on charges of fraudulently obtaining more than $8 million in federal COVID Relief funds. A friend of mine suggested that we shouldn't attend church due to this type of behavior by church leaders. I'll ask my friend Pastor Lalagofaatasi Sanonu for his perspective on this.

Washington Post reported during this period that Bishop George Davis, a Black pastor of the Impact Church in Jacksonville, preached to his congregation that getting vaccinated against the coronavirus was an act of faith. He said that he believes in divine creation, and that the shot is a miracle - a sign of God guiding scientists in their attempts to curb a devastating virus. Not all of his congregation agreed. Six of the church's members, later died of covid-19 in 10 days. Four of them were healthy and younger than

35; all were unvaccinated. The bishop later encouraged and organize a vaccination drive after each of Sunday's three services.

All over the world, the pandemic divided people across spiritual, ideological and racial lines, and the debate between science and religion was fierce. **COVID-19 is a science matter not a religious one**. According to some locals interviewed for this Book, the Samoan Island also had its share of people who refuse to get vaccinated. A few local anti-vaxers (who were definitely not scientists) even send out false claims on the internet. COVID-19 is a medical issue and people should listen to doctors (the experts) and not to anti-vax pastors (the stupid ones). Another illustration of stupidity is when a pastor died of a snake bite during a church service, while handling a rattlesnake and preaching that his faith tells him that a snake bite wouldn't hurt him. The misinterpretation of Mark 16:18 caused this event. So many idiots out there tarnishing the scriptures.

According to the Associated Press (AP), Facebook has rolled out a new prayer request feature, a tool embraced by some religious leaders as a cutting-edge way to engage the faithful online. Others are eyeing it warily as they weigh its usefulness against the privacy and security concerns they have with Facebook. In Facebook Groups employing the feature, members can use it to rally prayer power for upcoming job interviews, illnesses and other personal challenges big and small. After they create a post, other users can tap an "I prayed" button, respond with a "like" or other reaction, leave a comment or send a direct message. During the COVID-19 pandemic the new platform saw many faith and spirituality communities using this service. For better or worse, it's hard to argue that religion is big business these days according to AP. In a related story, A Texas pastor of Fort Worth's 2nd. Mile Church who had repeatedly questioned the COVID-19 restrictions lost both his parents within hours to coronavirus complications in 2021, and he seem to still feel he can't do anything differently. Stubborn atheist. The parents were parishioners of his church. Yes, he could have been less stupid with his sermons, according to a reviewer of my Book.

The global leadership group of The Church of Jesus Christ of Latter-day Saints did appeal to followers around the world to get vaccinated and use face masks in public meetings. Making the appeal amidst a rise in global COVID-19 infections, mainly driven by the Delta variant, the church's First Presidency made the appeal to their faithful: "We find ourselves fighting a war against the ravages of COVID-19 and its variants, an unrelenting pandemic," the First Presidency said in their appeal. "To limit exposure to these viruses, we urge the use of face masks in public meetings whenever social distancing is not possible. "To provide personal protection from such severe infections, we urge individuals to be vaccinated, available vaccines have proven to be both safe and effective".

God created man and gave the scientists the brains to implement related cures utilizing **existing ingredients and processes found in nature**, which he already created. Here are some examples:. Bone marrow has been suggested for a cure to AIDS. In early 2022 US scientists who have been working on the mRNA concept have accelerated research into "asking body cells to produce specific proteins" to fight COVID-19. COVID-19 vaccines used during the COVID-19 pandemic use the mRNA (already inside the human body) approach which is a concept that seem to be new but is actually an approach researched for decades. In early 2023, Asa Gustafsson, Professor of Pharmacy and Pharmaceutical Sciences, University of California, announced that "Cells routinely self-cannibalize to take out their trash, aiding in survival and disease prevention". In early 2023, the medical community confirmed that the cure for HIV has been discovered after an assessment of a patient they called "the Dusseldorf patient". The cure used the method of stem cells transplant. The plasma found in human blood (mostly donated!) has been used extensively by Pharmaceutical Companies around the world to manufacture drugs[42] that have saved hundreds of lives. I think God really created his "masterpiece" with everything it needed

---

[42] Jave Discover. Free G. "Harvesting the Blood of America's Poor".

to stay well. Nature has also provided blueprints for some recent innovations like the turbine technology where entrepreneur Frank Fish (Philp Watts) observed bumps on whale flippers and used the structural concept of "increasing the lift and simultaneously reduce drag" to improve blades efficiency in pumps, planes, wind turbines, fans compressors and other items.

Local cures for many traditional ailments by traditional healers (*Taulasea*), utilize local plants and other native resources. In 2022, The University of Queensland[43] announced that it has found out that "superworms" feast on plastic waste. The serious problem of trash/plastic has now begun to see the light. In Paraguay, the water dams provide about 90% of the electricity. About 14 nuclear plants and the equivalent of thousands of gallons of oil per day would be needed to produce this amount of electricity. With relatively minor sacrifices, the long-term effects would mean much less people getting cancer from nuclear plants, significantly less air pollution, and several overall positive impacts to the wellbeing of the community and the environment, all because of the Christian God's water resource. What a Creator.

The Christian God created a perfect planet Earth (Genesis 1:31) with its own solutions to its various problems. Unfortunately, humans have continuously contaminated the environment, and caused diseases; then God send down many more scientists to solve these problems but the spoiled brats (so called human beings) killed off these experts through abortion, World Wars, invasions of other countries (like the invasion of Ukraine by Russia during this period), and miscellaneous domestic and foreign killings. Way to go humans!. Furthermore, humans continue to celebrate and glorify the killings of other humans!. For example, Palestinians in Australia, Britain, the US and other countries made notable celebrations on the streets when Israel was attacked by the terrorists group Hamas in 2023. (This terrorist group also destroyed the world's third oldest church).

43

Many local residents of these communities said that the protesters needed to be send back to their countries to hold their protests at their backyards. The numerous protests in the US streets may have triggered the Republicans (Rep. Ryan Zinke) to introduce a bill to expel Palestinians from the United States.[44]. Fox News reported in November 2023 that the current rampage criminal activities in Sweden are mostly caused by immigrants, but then this is just a very small group of rascals.

Noisy, troublesome, rowdy and disruptive protests are not welcome in my village, especially during our village evening 'sa" (curfew) and I understand Ryan's concern. Interesting though, there was a peaceful march planned for December 6, 2023 in Samoa, petitioning the government for support of a Middle East cease fire. It was mentioned at this time that the Samoa government did not vote in the UNGA on October 26, that called for a resolution for an immediate truce between Israel and Hamas. I personally hope the celebrations of the killings of innocent private citizens won't happen in these islands because it would be unsafe since most Samoans are pro-Israel. According to the Pew Center, there are an estimated 7.5 million people in the US that identify themselves as Jewish and about 3.5 million as muslims. Some comments in social media around this period suggested that the US has become the world center of mass shooting. A kind of pandemic itself.

---

[44] Delaney, Arthur, Abdelaziz. Nov. 3, 2023.

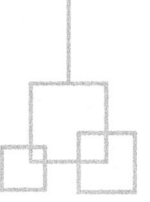

Abortion

This controversial issue has been a weighty topic in secular and religious discussions for many decades around the world and also in these islands. I think this issue is a women's personal issue and decision. A women's body is her personal business not a church nor a government property. Other issues related to this topic is the idea that Mormonism and muslim sometimes allow multiple wives and when an Arab woman is raped, honor killing comes into the picture. Personal choice and human rights must take precedent in this era. Not politics. If pondered in the religious arena, well, get ready for copious and varying perspectives.

In September 2021, Pope Francis said aboard the Papal Plane that abortion is "murder", even soon after conception. He also criticizes US Catholic bishops for dealing with US President Biden's pro-choice position in a political manner rather that a pastoral way (Reuters 2021).

The General Secretary of the Congregational Christian Church of Samoa (CCCS.) believes that the biggest denomination in Samoa might "generally" reject a call to legalize abortion in Samoa. Reverend Vavatau Taufao said that personally, he would not agree to formulate a law to legalese abortion in Samoa. "I'm very careful as I am not speaking on behalf of the church," he said.

"The church hasn't made a stand on the matter, and I don't think the church accepts a law that would accept abortion generally."

However, that should not stop conversations and discussions of the matter, said Reverend Vavatau. "The question of whether or not we should consider it, I personally believe that we should talk about it and discuss the issue. He continued: "Because there are factors that we need to consider when talking about abortion, and there is a recommendation for it to be made legal in Samoa, then we need to consider all factors like medical factors, when it is a matter of life and death". He added: "If abortion saves a life, then why not". "But when it comes to other cases of rape, I cannot accept a law that will enforce and encourage the girls or female that once you get pregnant because of rape, then you must immediately abort the baby". I think the decision needs to come from the victim. So, there are factors we need to consider when it comes to abortion, but generally, I don't think local churches will consider legalizing abortion, however, it shouldn't stop us from talking and discussing the issue". He concluded: "As I said, there are things to take into consideration and that can only be done through discussions.". I agree, but the issue has been discussed for many years and I think there would never be a consensus.

The recommendation to legalize abortion was part of a submission by the Unites Nations (UN) country office as part of the Universal Periodic Review (UPR.) to be undertaken by the UN Human Rights Council. Samoa is among 14 states whose human rights records will go under scrutiny with the Council's session to be held in November during this period. Under the Crimes Act 2013 the only exception to criminal liability is if the pregnancy is terminated to preserve the woman's life or her physical or mental health and within the first 20 weeks of gestation. In 2010 the first case where a person went on trial for procuring abortion emerged in Samoa with a former nurse who illegally oversaw the termination, was jailed for four years. The UN Committee on the Elimination of Discrimination against Women, in its submission to the UN, raised concerns about Samoa's limited grounds for legal abortion. It suggested this be considered in "at least in cases of rape, incest, severe fetal impairment and risk to the health or life of the pregnant woman, and decriminalize abortion in all other cases". This to me seems to be a good start.

In 2017 a similar call was made by the Ministry of Health in the National HIV., AIDS. and STI Policy 2017 – 2022 report. At the time the report noted the abortion service is critical to the health and wellbeing of people living with HIV, those dealing with sexually transmitted infection and survivors of rape and incest offences. The recommendation to legalize abortion in the country has also been rejected by the leadership of the Opposition party, the Human Rights Protection Party (HRPP). Former Prime Minister Tuilaepa had said that Samoa should slam the door shut on calls to legalize abortion in Samoa and that abortion is murder. I think he should talk with victims of rape, and incest because government laws should consider everyone's circumstances especially the victims' rights and he shouldn't sway towards **his Catholic** conviction, like the current US Supreme Court's decisions now ruled by the Catholic majority.

In 2021 (according to the New York Time Company) the U.S. Supreme Court failed to rule on an emergency application on a controversial Texas law that bans abortions after six weeks of pregnancy. As such, the legislation went into effect on Sept. 1, 2021. Some people believe that millions of children lose their right to live their lives because of abortion. The catechism of the Catholic Church states that "Since the first century the Church has affirmed the moral evil of every procured abortion". This is however, only one part of the history of this subject. In the Letters of Apostle Paul, it discouraged marriage and reproduction. Later Christian texts supported these teachings. In a second-century text known as the *Acts of Paul and Thelka*, a Christian author told of Thelka rejecting her suitors in favor of spreading Christianity. Thelka's story inspired Eugenia, a Roman nobleman who rejected marriage (*Acts and Martyrdom of Eugenia*) and led a male monastery for a time. Several men in Rome convinced Emperor Gallienus that Eugenia's teaching will endanger Rome's military power since it reduces the number of future soldiers. Eugenia was executed in the year 258. In 211 Roman

Emperors Caracalla and Septimius Severus allegedly made abortion illegal. Some bishops sometimes condemned the injustice of laws regulating sex and reproduction. For example, the bishop Gregorios of Nazianzos, who died in 390 accused legislators of self-serving hypocrisy, for being lenient on men and tough on women. Similarly, the bishop of Constantinople, Ioannes Chrysostomos, who died in 407 blamed men for putting women in difficult situations that lead to abortions.

There was an increase in rape cases in these islands during this period. Politics play an important role in government's resolve in this matter. HRPP's leader is catholic and a staunch supporter of making abortion illegal in Samoa. The fact is that Christians who support women's reproductive rights are also following the historical precedent of their religious tradition. Government regulations should stop at the door of peoples' homes and should never enter the family home, especially the bedroom.

Guatemala, in early 2022, increased prison sentences for women who have abortions.

During this period, Prime Minister Fiame Naomi Mataafa clarified that her government will not change the laws only to satisfy some of the recommendations made by the United Nation's recent Human Rights Report. Some of the fundamental rights central to the UN report included rights to abortion, same sex marriages and euthanasia. These issues were raised in the Universal Periodic Review of Samoa's state of Human Rights as examined by the United Nations Human Rights Council (UNHRC) earlier. The Prime Minister Fiame emphasized that these are fundamental rights, which are well-protected under Samoa's Constitution and that whatever issues the UN raises, the Government's utmost priority is the local environment and the Constitution and the context of those issues within the Samoan traditions and customs and also our Christian principles and values.

Related to this issue is the controversial topic of killing another human being. Various perspectives relating to this issue exists all around the world and a myriad of circumstances requires a careful assessment of each case. A 2023 example mentions a 77-year-old woman from Florida who was accused of fatally shooting her terminally ill husband after they agreed on a suicide pact due to the unremitting effects of the man's illness. Evidences were available supporting the suicide pact; however, the woman was jailed. The woman was supposed to kill herself after she shoots her husband but she couldn't do it. I'm one of those people who would refuse to live my later years in a vegetative state and would opt for a hospital setting where they would administer euthanasia in a legal way. In recent years, several people have been legally euthanized in the Netherlands – the first country to allow euthanasia in 2002. Again, any reader is welcomed to debate me on this subject but must visit the homes of the elderlies, Hope House, *Mapuifagalele* and hospital ICU, and all the terminally ill people before approaching me. I don't allow theorists to debate me on this topic, only **realists** are allowed. We're not living in a fantasy realm bubble but in a real world on planet Earth, a real planet. The terminally person who asked to be let go and who is in really serious pain 24/7 gets to decide. Not the healthy relatives, doctors and friends and definitely not the legal system. What do they know. A very sick 5-year-old girl name Juliana Snow was asked by her parents to decide if she wants to return to the hospital to spend her last days and suffer continuous pain from several treatments or stay home with her family and she opted to stay home. Get educated people!

I personally know and experience this phenomenon when my wife would call me at 2 AM in the morning from the hospital saying she can't take the pain anymore. She died a few weeks later. Additionally, I had a similar experience with my eldest sister, Initia, during her last three weeks on Earth, in Arizona. During those three weeks, I would softly rub her hands for hours and she would talk to me about the **unbearable pain**. Similar instances happen to my nephew and brother. The pain was so horrible, they wanted to leave this world. So, I have a ton of real-life experiences on this subject and **know exactly what I'm writing about**. Getting knowledge only from text books and questionable so-called experts, is seriously insufficient to form an opinion. It's also very stupid. Jesus's teachings mainly used real life situations with spiritual advice. Physical pain is not spiritual (Christianity discretely applies to the human spirit not the physical being). If one kills the physical body the spirit (which is more important to Christ) lives on.

This brings up my continuous disagreement with the reading of the Bible during local funerals, where pastors would usually read 1 Corinthians 15:55 "Where, O death, is your victory? Where, O death is your sting?. Atheists and many baby Christians attending would be confused as to how the faith has won, yet the deceased has clearly died and **death has won**. I humbly suggest that these pastors should first read John11:26 'and whoever lives by believing in me will never die", and then 1 Corinthians. This arrangement would elucidate the related scriptures regarding the Christian view on death and lessen the confusion. Christians should make every attempt to convince atheists that we are preaching the truth by clearly laying out scriptures in an orderly manner. Stop helping atheists make fun of your faith.

I consider myself a realist but also a baby-Christian, and these two don't blend because conflicts exist between their two basic principles. God didn't create man in flesh with a spirit at the SAME time. Flesh was created and then He breathed life into man afterwards. There are scores of scientists and professors who are theist-scientists but are struggling to reconcile these two ideas. I don't think that is possible. One can still live as a scientist and also attend church as a good Christian. As I will explain later, Christianity is purely a spiritual phenomenon (Jesus also said that) and the physical human experience is separate and therefore should be portrayed in a different realm of reality. Talking about being a realist. I attended a church service in our AOG Pago Pago church in early August 2023 where the pastor said that God will protect everyone whether they are at work, going to school or spreading the

Gospel of Jesus Christ. I immediately looked towards my right where a woman whose husband recently died while at work sat quietly. I was confused about the pastor's message that morning service and wondered what the poor women was thinking.

Sunday school teachers and Bible College instructors who find it hard to explain the concepts of "loyalty and devotion to Christ" may use a secular phenomenon that actually occurred in 2023, where scores of people went to jail in the US because they believed in a personal conviction of only one man in relation to the Presidential elections.

Hundreds of faithful Christians get cancer (like my wife) and after hundreds of prayers, they finally pass away because their illness is physical, but I know their spirits live on. Dozens of parents in the US have refused to take their sick children to the hospital but preferred to leave them at home and depend on prayer. Many children consequently died from this practice and dozens of parents were jailed. Reality and faith sometimes just don't socialize, and novice interpretations can confuse the faithful.

Readers can use the following statistic to argue their opinions from either side of this issue: According to one recent study[45], around 64,000 women and girls become pregnant due to rape in US States with abortion bans.

---

[45] NBC News. JAMA Internal Medicine. Megan Lebwitz-1/24/24.

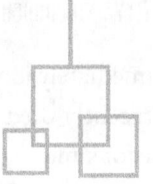

Evil and the devil

L et me explain the idea of evil. Some people categorize evil as:

1. Moral evil. For example, the 9-11 attack, the recent invasion (not a war) of Ukraine and the NAZI's killing of thousands of people (many were children) in gas chambers.
2. Natural evil. For example, the Ebola pandemic.
3. Bigotry-supremacist evil. For example, a Jewish non-verbal teenager with autism returned from school with a swastika carved into his back in Nevada in April 2023.

It is a good idea to educate people on this topic so they are made aware of the many devious ways Satan can intrude into the every-day lives of our island community. Knowing one's enemy is crucial in attacking and defending during wars. The devil phenomenon is real. In fact, Pope Paul IV once famously said "the devil is real, not a hype piece for the movie" and "one of the most important things we need church for is the protection from the devil". The islanders are in a spiritual war according to our local pastors. Additionally, these islands are not removed from the rest of the world. Our economy depends largely on exports and our social conversations and contemporary perspectives are seriously influenced by television and social media. With this plainness, devilish behaviors had invaded the layback island-style life and therefore the locals should be well-informed on this topic. In March 2022 as I was writing this book, television news reported children, pregnant women, hospitals and even churches and mosques being bombed by Russian troops. "This was the bloodiest regime" wrote Gorky Siemaszko (NBC News 3/18/22). Who would bomb churches and mosques (BBC. Jack H 3/30/22) but the devil's friends. I'm not the only person in the Pacific interested in this topic. A new book "The Devil Exists" by Oliver Peres of New Caledonia was published in 2022.

There is a misconception that evil acts always result in bloodshed. It is far from the truth. Evil acts also occur in White Collar dealings and in most cases have far more devastating effects on relatively more innocent people than crimes that physically hurt people. For example, thousands of elderly people in the US are routinely shattered by the loss of their lifetime earnings through various forms of financial scams. For example, Ponzi Schemes like those perpetuated by the infamous Madoff[46] during this period. These schemes are not new. In fact there was a case in 1872 relating to an inventor, John Keeley from Pennsylvania who duped many investors with his fraudulent inventions. A US company also tried to get investors for its fraudulent concept of trying to use energy in empty space, which according to some top Theoretical Physicists, is impossible. Readers can read more on this subject in Robert Park's book "Voodoo Science".

These scams also occurred in the Pacific. In 2023, it was reported that the US Federal Bureau of Investigation (FBI) was seeking to identify possible victims of a fraud scheme that was allegedly operated by Tilila Siola'a Walker Sumchai. This scheme allegedly defrauded the Tongan community out of $13 million and the US Securities and Exchange Commission (SEC) has also file its own related but separate complaint. The Tongan community defrauded are mostly, if not all, Christians from various denominations.

Recently though, a group of "anti-scammers" came together and formed what they call the "Peoples Call Center" where the

---

46

group stopped dozens of evil scammers world-wide and planned to continue their mission periodically. Members of this group had stated that the are "always there to protect people" and they can be reached at scammerpaybacvktipline@gmail.com. This group can be likened to those that stood up against the deceptive US January 6 insurrectionists and I hope businesses and the public would donate to this type of initiative since it aims also reflects Biblical teachings regarding helping the poor.

Furthermore, millions of ordinary people often pay for the unscrupulous manipulations of financial information. For example, about $250 billion per year in taxes are not paid by mostly wealthy people and corporations, mainly through Tax Shelters. This means that ordinary people should have doubled their tax refunds if these evil people would have paid their legal share. This is according to former (1997-2002) IRS Commissioner Charles Rossotti. An interesting example of tax evasion is the concept of LILA -Lease in lease out, where a German company leases street cars to a US company then immediately leases it back. The losers in this type of arrangement are often the ordinary people of any community, including many poor Christians.

Evil has also made churches and mosques unsafe. In 2023, a Minnesota man was charged with second-degree murder after fatally stabbing his wife during a Bible study session[47]and several people were killed in the US while attending church services and Bible studies.

The devil or evil works can be compared to the so called "Red Herring defense" which was used in US courts. The Red Herring distracts predators by moving fast in front of them to try to save the targets. The devil will always try to distract people from the truth by throwing in numerous deceitful "theology" and misguided perspectives to hide the truth about the Christian God. The most ridiculous revelation regarding this deceptiveness is that many of the

---

47

cunning people involved in distracting parishioners from the truth are so called preachers, notably televangelists of Mega Churches. Notable examples are mentioned in another Chapter.

In case a curious person hopes to have a glimpse of the devil. Well, he was/is here on Earth. His personality was mirrored in people like Adolf Hitler, Saddam Hussein, Nazi Adolf Eichmann, the hijackers of 9/11; Dr. Josef Mengele, Larry Nassar, serial killers, mobsters and murderers all over the world; prosecutors and police who framed innocent people; terrorists and company executives whose goal is making profits regardless of the poison they inject into Mother Nature (e.g. Du Pont case) masterminds and affiliates of Ponzi Schemes; Jeffrey Epstein; but more damning are the preachers and cult leaders - those who lead people into their cults and eventually their deaths and also those televangelist who milk millions of dollars from the disadvantaged population to buy luxury homes, jets and other lavish items for personal use!. According to TV personality Jon Stewart these types of human beings are "comic, absurd, shameless and do shameless things" (Huffpost. Sept 11. 2021). This author believe that the devil will never be defeated on Earth because humans are willing hosts; implementing the devil's commands to perform atrocious acts on Earth. But I think, most of the time it's the hosts themselves who actualize these evil acts **on their own**. Some unwillingly like those with mental illness. The devil is like a virus. It needs a host (humans) to survive. Therefore, if humans refuse to accommodate the devil, then it will eventually expire.

What about justice against any evil act?. The US Legal System responded to a mass killing in New York in April 12, 2022 through Judge William Kuntz II announcement which he said that "perfect justice would require a power that neither this judge nor any other judge has in his or her hands to impose". I think that he was referring to or relying on the power of the "Last Judgement" mentioned in the Bible, to provide the final justice since he, and other judges, wouldn't be able to provide a suitable sentence for this unspeakable crime?.

Evil may also manifest itself in a more subtle way where it comes to destroy families (John 10:10), like parents who didn't get vaccinated for the Covid19, died, leaving many young children in very traumatized and extreme situations. It may also manifest itself in bewildering circumstances; not through perpetrators but, for example, through a judge who sentenced a man who shot his pregnant wife ten times to **only** 5-15 years. For people living overseas, it can be surprising and chilling that through advances in forensic science, investigators can now identify friends, associates or even family members that are/were murderess. The devil maybe lying next to you!.

In the Samoan islands, it had manifested itself through politics where the defeated political party refuse to follow the law, resulting in several breadwinners being terminated from their jobs, leaving a few families with little or no food. Scams that promised overseas seasonal jobs in New Zealand around September 2021 in Upolu were also the work of the devil. Moreover, a few foreigners were murdered by local males on the island of Upolu in the last decade and a Vietnamese male and a local girl were stabbed on Tutuila in October 2021. In New Zealand, Joseph Auga Matamata aka Viliamu Samu was sentenced to prison for human trafficking and enslaving 13 people (RNZ Pacific 4/12/22). Other local reflections of related evil influences include the significant increase in the abuse and domestic violence against women and children, as reported by local government statistics in the past 20 years. I believe that a major contributor to this situation is the increase in drugs use as reported daily by local media.[48]. In late 2022 in American Samoa, a young man (Seilala Onofia, a friend of the devil) was convicted of stealing equipment belonging to his church and was ordered to serve 28 months in jail.[49]

Some people have used the phrase *ua ao Samoa*, meaning Samoa

---

[48]

[49]

has been tamed (or civilized) through Christianity. Yet, there is little room in prisons and additional new prisons need to be built to accommodate the significant number of offenders. In Samoa, there is a significant increase in the number of abused women and children sheltered by the Samoa Victims Support Group (SVSG) during this period. Some local pastors like Pastor Tino Sauaga and others who were featured in Daniel Pouesi's video presentation "In Search of *Tagaloaalagi*", disagree with said saying. Generally, this phrase has some truth in it. Unfortunately, many offenders are lost souls, some failing to find jobs or other ways to move forward in life, and some have been manipulated by the ever-increasing drugs flowing into the islands during this period. The phrase should therefore be corrected to: *Toeitiiti ao Samoa*. Like in many countries, a myriad of factors come into play and need to be addressed before these islands can progress into the *"ao"* stage.

It is preposterous to read about the horrendous things humans do to other humans. As I mentioned in my previous book, the horrible, hideous and atrocious acts done by vile humans to their fellow friends, relatives, children and everyone else truly makes the devil look **like a Sunday School teacher**. A very disgusting example that stands out (besides the atrocities done by the Nazis) is that of a man name El Pozolevo, who made human soup for the cartels, out of human flesh. The idea that mankind was created in the image of God has always baffled me in relation to evil acts by humans, but I think the "image" refers to the comparison of the Trinity to the humans' three identities: Body, soul and spirit. It has nothing to do with the evil acts of humans which is from their own decisions and freewill, and the rest from mental illness. Some scholars have suggested that human beings should not be classified by just the chromosomes but by their actions and convictions. According to this definition, jihadist who behead people are not humans but alligators. I was not surprised when I talked to a few locals and they agree with this classification.

According to some pastors the devil normally "suggests" but

it is the humans who actually execute the sin. My friend Frederick Obrien (a former US JAG Judge, Acting Governor and Attorney General) agreed that our family dogs are better living things than these cruel perpetrators. A famous actress Brooke Shields (we were both born May 31) noted in the TV program ET in March 2022 that her relationship with animals was much better than with humans. No wonder a few wealthy people in the US left thousands of dollars for their pets in their wills. In 2022, a dog name Sarah saved a wheelchair user from drowning (Treasure Coast Newspapers. W Greenlee 2.4.22) while forensic scientists recently concluded that the skeletal remains of two boys that were discovered in Vancouvers Stanley Park in 1953, indicated that the boys were bludgeoned in their heads with a hatchet (Oxygen True Crime. Jax Miller 2-16-22). Animals vs humans. A serial killer started killing homeless people sleeping on the streets of New York in early 2022. Who would intentionally harm these unfortunate souls except the devil's friends. Interesting though, the Bible also has several very sinister accounts of evil events mentioned in the Old Testament.

During this period, in a sense, pigs are even more useful than some of these murderers, abusers and rapists. According to USA Today (1-20-22), Birmingham and New York hospitals have started implanting pig kidneys and pig hearts into some patients. The waiting list for these organs in the US is about 100,000 and these animal organs will eventually save lives compared to humans killing other humans. Of course, there is one "mitigating" factor in some (obviously not all) of these cases. Mental illness, where people with mental issues tend to kill indiscriminately.

Samoans in the past also had their share of atrocities against their own people as described by Dr. Augustin Kramer in his Book "The Samoa Islands"[50]. Most of these, however were activities done before Christianity was introduced into these islands in 1830. Deliberate evil acts mentioned above, on the other hand, were performed many

---

[50] Translated by Dr.Theodore Verhaaren. Volume II. pp 185

years after Christianity was introduced to the US public and is still continuing as of today.

I've been struggling with this issue of where the depraved events started and who is responsible for these merciless activities that happen throughout the world. After a few years of trying to resolve these issues personally, I finally figured out the answer. It was right there in Genesis, starting at the Garden of Eden. Man (Cain) started all this, not the devil, as many pastors have alluded to. The devil was busy in another part of the garden of Eden. So modern day humans are responsible, including mentally retarded people, for all the atrocities on Earth. Readers can watch numerous well documented series (not movies) on the internet regarding court cases on murders to prove my point and be fore-warned one might just throw-up (literally as I did) after viewing these unbelievable crimes committed by humans, **not Satan**. God didn't create these monsters; it was through the conjoint chance union of their parents and the effects of the "environment"[51] according to Pastor Iliafi Esera. I honestly believe that the devil has been hiding for many decades, and **very afraid of humans**, especially the serial killers who decapitate women and children; those who shoot young children in the face; those who boil human parts inside their homes, stab elderly people and kill the homeless. These acts are all well documented in the many documentaries on the internet, related books and articles worldwide.

In case Christians think that satanic worshipers is just an unorganized group of ludicrous and irrational humans, they should know that some of these smart people are reasonable humans, have their own believes, temples and don't believe in the Biblical Satan. For example, there's a fine Satanic Temple of Illinois under the direction of Minister Adam. As for atheists, I personally know some of these people and they are good moral and ethical humans, sometimes much better creatures than many locals I see in churches

---

51

on Sundays. Morals can exist without religion, it's a fact, and its already in human DNA created by the Christian God.

Around Easter 2023, a mother and her children were saved by the Police and the SVSG in Samoa, from the prolonged abuse by the evil husband, stepfather and grandfather. The evidences and testimonies revealed numerous wicked acts experienced by the poor mother and her children. I don't believe in the Muslim rules against abusers and thieves, but this is the kind of case where I believe the Muslim laws should be applied in these islands in addition to normal laws of the land. One might argue that these abusers have rights and should be treated fairly. According to one of my *tulafales*, abusers of children and women have no rights. The churches should do more outreach regarding this problem instead of just "praising the Lord" in the comfort of their churches. Get out to the "local Gentiles" my friends and do some actual work for the people of these islands.

When I wrote "just a little" help from the devil, as the perpetrator of evils events, I was in part expounding on court cases where culprits use drugs like LSD and causes them to act out in very heinous ways. In these cases, it's the drugs NOT the devil that caused horror. Most of the time, humans do more horrific acts than those where the devil is accused to be the enabler. But readers don't have to believe me. Do some research on the web and watch the numerous related case studies, court cases and documentaries available on the internet. It is interesting to point out that there's little differentiation between humans and apes in that the human genome is about 99% identical to a chimpanzee, according to some researches. I really hated this comparison until I started doing research on serial killers, pedophiles, genocide instigators and the Nazis. This is when I concluded that these killers are actually animals and are so evil that even the devil himself would be scared of them. The scientific statement that human genes are very similar to those of animals like chimpanzees initially seem to be very revolting to me. However, the above observations corroborate this science fact.

Genocide is also not an issue from way back. In the early 1900s, there was a genocide in German Southwest Africa by the German force *Schutztruppe*. Read about this incident and one would be astonished about the related evil events.

For the mathematicians, Evil can be described as the numbers on the left of zero, where everything is "negative". Every "goodness" in mankind can be represented by the numbers on the right of zero, where everything is "positive". That leaves Mr. Zero which can be interpreted as God, the beginning of everything good and his "allowing of man's freewill" to start everything evil.

Some evil acts happened in the pacific area during this period. For example, a mother was arrested in Papua New Guinea in 2022 after allegedly killing her three children and dumping their bodies in a river. I'm mentioning this activity because my parents were London Missionary Society (LMS) missionaries in Papua New Guinea before I was born.

A 2023 case in American Samoa portrays an example of how alcohol affects a person's behavior, where a male[52] reportedly set fire to a pastor's house in Petesa. I don't think the devil had anything to do with this event and police concluded that it was alcohol-related.

An example of modern-day evil is the cyberattacks on US hospital computer systems where thousands of patients' lives were affected. An angry relative of one patient said that these hackers should be executed. Another modern-day example of evil happened when Hamas kidnapped a young girl with special needs and her grandmother then shot them, this is in addition to killing the parents of a 4-year-old in front of this child then kidnapping the parents. I propose that even the devil himself can't be this evil. The numerous discussions that followed on national television and social media failed to mention past incidents like the 1941 Fahud program, the Sol Hachuel incident in Morocco, the Machhad 1839 incident and many others[53].

52

[53] Travelinisrael.com "The ethnic cleansing of Jews within Muslim countries".

Abhorrent crimes against other human beings are all disgusting to say the least but mystery and crime writers continue writing and selling thousands of books describing these horrendous acts. The irony is that these writers don't want these atrocious acts to end so that they can continue to reap the fruits of their labor!. Its not their fault these events occur during their lifetime, and they might as well utilize their talents and make some money.

Comparing the powers of the devil and the spirit of Christianity has been experienced by several communities in the past decades, where dedicated Christian parishioners have been violated by the spirits of the dead, including local students studying at local Bible Colleges, e.g. at Kanana Fou (Niuatoa 2007).

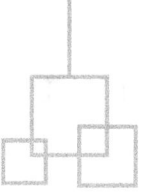

Related Miscellaneous Developments

In 2021-2022 there was an increase in the number of people waving signs near the main roads in American Samoa proclaiming that "Jesus is coming soon". Predicting when Jesus will return has been a farce for many years in the Samoan islands especially by the Seventh Day Adventist Church in the past. I personally attended some of these programs at Lalovaea, Samoa. These silly campaigns should cease **because even Jesus Christ himself didn't know exactly when he will return** and because these predictions have always been completely wrong so many times for hundreds of years, it had consequently discouraged many possible recruits to Christianity. One of the devil's tasks is to redirect Christians away from the Gospel of Jesus Christ and it has been suggested by some locals interviewed for this book, that these campaigns should cease as it is the devil's work. Have these people read Revelation 22?.

In the US, televangelist Pat Robertson made related ridiculous predictions about the end of the world (for 1982 and later changed to 2007 - HuffPost). Maybe he is a friend of the devil according to one local chief. Unfortunately, he passed away in June 2023 without reading this book which may have made him repent and go on to everlasting life… maybe.

In the Pacific, Fijian authorities captured and deported two members of the Korean doomsday Chistian cult, Grace Road during this period, while the group's leader Daniel Kim and others remained on the run.

In Washington, in front of a private social club, stands a statue honoring the 19th century US Navy sailor John R. Monaghan. Many residents have no idea who Monaghan was. But for many Pacific Islanders in the area, his legacy is one of violence, colonialism and racism — and they want the statue to come down. Pacific Islanders and their allies marched through downtown Spokane during this period to amplify calls to remove the monument — a demand that community members have made for years and have since renewed recently. Joseph Seia, founder and executive director of the Pacific Islander Community Association of Washington is at the forefront of this issue and related activists have started a petition to take down the memorial. "There's no honor in lifting up somebody that killed our ancestors," Seia said. "Monaghan is believed to be the first Washington resident to attend the United States Naval Academy and served in the Navy at the height of US imperialism in the 1890s", according to historian Lawrence Cebul. In 1899, the Navy attacked Samoa in an attempt to impose US rule over the islands. The USS Philadelphia — on which Monaghan served — shelled native villages, while officers went ashore to burn survivors, Cebula said. Their actions targeted and killed civilians, including women and children. The statue of Monaghan in Spokane was erected in 1906 in a ceremony that included numerous racial slurs against the Samoan people, according to Cebula. Those slurs and offensive characterizations persist to this day — a plaque at the base of the monument refers to the Samoan forces as the "savage foe" and inaccurately depicts them with bows and arrows. The New Zealand government recently made an official apology to the people of Samoa regarding the Mau movement and there is an old *Vaimea* prison in Upolu where prisoners were allegedly hung during the Mau Movement. Speaking of statues, the Deacon of the Catholic Church at Leauva'a-Uta had expressed his disappointment and sorrow at the demolition of the statue of the Holy Mother that stood by their chapel. The Deacon said he was saddened by the way the statue was bulldozed before his church had time to remove it themselves. This event stemmed from a court case relating to a land dispute between Afega and Leauavaa villages.

In the world of politics; the most powerful message that has been successfully used for many decades by ruthless and indifferent rulers is that certain races are far more superior than others. This was used by Hitler, Mussolini, a few Presidents. Several hate groups in some countries utilize the same race supremacy hypothesis and some powerful countries had also used a water-down version in the administration of their colonies around the world including the pacific island nations. The white nationalists and supremacists in the US have ignored the fact that it's the native American Indians who should advocate these nationalist creeds because America is their native land and they (modern white supremacists and nationalists) are actually **foreigners (or aliens)** who stole lands from the Native Indians. (The famous astrophysicist Neil Tyson has also said that "white people look more like monkeys than Black people"). A misplaced and therefore ridiculous dogma. Carlin, a notable standup US comedian noted that the white people of the US "stole the lands from the Indians and Mexicans". So, what's the basis of these nationalist and supremacist's convictions?. In fact, many members of these groups were jailed due to their related actions during the January 6 insurrection events in the US capitol, proving that their convictions were untrue, misguided and illegal.

As churches increased its use of modern-day music technology in their church services and various youth programs, the traditional dances and songs performed by large village groups and organizations at national events followed suit with the use of contemporary keyboards. These were not used in the American Samoa national performances up to the early 1980s. The yielding natural sound of large groups singing in the past genuinely portrays the cultural messages reminiscing the past and hinting on needed changes for the future, all without modern instruments. The western-style sound from these keyboards especially when using their distort functions were un-Samoan, offensive mechanical noises, and downright ridiculous. In fact, the late Director of the American Samoa Arts Council, Leala Hanipale Pili (late wife of an LMS pastor), who supervised annual national cultural programs featuring traditional dances and songs for many years, banned the use of keyboards in national holiday events including Flag Day events before she retired. Late icon traditional Samoan musicians Ueta Solomona and Palauni Tuiasosopo also agreed with me that the keyboard noise from these contemporary instruments were not appropriate for these national events which was supposed to highlight only **traditional** performances and songs. As a professional musician and former Music Studio Producer, I totally agree with the above-mentioned experts, and personal friends. Church music is a different being but has infused cultural mentions in its lyrics lately. For example, a Catholic church on Tutuila included Samoan cultural sayings in a couple of hymns shown on KVZK TV during this decade. Overseas, a similar phenomenon occurred during 2022 when the US Army military band added Rap music to its usual music selection for public performances. Music and the church.

In June 2022, a pastor (Tavita Kapeli) of a Falelauniu church and people of that community displayed various home-grown agricultural products ranging from vegetables, fruits, root crops and ornamental plants. The Ministry of Agriculture and Fisheries provided equipment, farming tools, seedlings and various crops to this initiative. Most of the products were later donated to the *Mapuifagalele* and the Carmelite Sisters. This is a fine example of how churches have contributed to the betterment of village communities.

The Samoa National Council of Churches selected Rev. Aisoli Iuli as its new Chairman in 2022 after Deacon Leaupepe Kasiano completed his term. The late Rev. Oka Fauolo served in this position for many years in the past. This Council was established in 1961 and was setup to develop sincere fellowship and cooperation among member churches and to work together towards true Christianity.

Unfortunately, other churches like AOG didn't like this idea and didn't join the Council in the past. The Council is a member of the Pacific Council of Churches within the World Council of Churches.

The Samoan Heritage Week continued in 2022 where former Governor Togiola thanked church leaders for moving to revive this event. The event was to emphasize the importance of instilling a sense of pride in the culture and language of Samoa among the diaspora, especially the youth.

The London Missionary Society of Britain has a lot to do with Christianity in these islands and it was with a heavy heart that I paused for a few minutes, while writing this book, when I heard that Queen Elizabeth passed away in September 2022. I read about the Queen as a student at Samoa College, in Apia, in the early 70's and I also had a couple of teachers from Britain at that time. The relationship between Samoa and Britain was reflected in the Samoa Head of State travelling to Britain for the Queen's funeral. American Samoa also send formal condolences to that nation at that time. From the Chiefs of *Salani*, Rest in Peace Dear Queen.

In 2023, Pastor Warren Retzlaff and his wife Julie of New Zealand visited these islands to host a 3-day seminar series with Word of Life Church. The 3-day event was aimed at helping couples enhance their marriages through a faith-based approach. The event was called "Love Amplified".

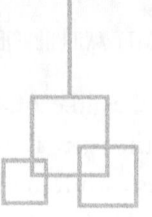

Related Topics

C hurch opposition to the former Samoa government's move against taxing pastors would have been much stronger if all the denominations agree and united against the new laws. Pooling together resources always provide a much stauncher stance against any issue. For example, twenty *fautasi* (traditional long boats) filled with warriors from *Savaii (Pule)* paddled to *Upolu (Tumua)* in December 1908 to greet the German Governor Wilhelm Solf at Mulinu'u. The purpose of their journey was to present their list of petitions to the German Governor and to hopefully gain support from the *matais* of *Tumua (Upolu)*.

An interesting quantum mechanics-related experiment was explained by scientist Michio Kaku where a cat was supposedly placed in a box with some poison, a counter and an atom of radioactive uranium. If the uranium decays, it sets off the counter which then releases the poison and should silently kill the cat. Before opening the box, the scientist said that we can't actually know whether the uranium decayed or not. He explained that the human conscious determines if the cat is dead or alive, and then he asked this weird question: how do I know if I'm dead or alive?. after watching this episode, I took a break and walked over to talk to a farmer (Lolo) who was planting some *taros* and asked for his view in regards to this weird question, and here is what he said. If the scientist asked him this question while standing right in front of him, he will punch his mouth so hard that some of his teeth would fall out, and if he's dead, he won't feel a thing but since he's alive he would feel pain and maybe ask the chiefs for forgiveness. Simple, he said. While we're on the subject of quantum mechanics, a comment was made on a podcast that 'science has not buried God, it has revealed Him, and with it buried materialism".[54]

US-made Javelin antitank missiles were dubbed "Saint Javelin" in a meme that circulated during the 2022 invasion of Ukraine, when these weapons wreaked havoc on the Russian invaders. The meme showed Mary Magdalene cradling a Javelin in her arms. Christian-related emblems can never prevent the coward intrusions by the devil.

An initiative called SHOAH (Hebrew for Holocaust) where interviews with survivors are stored in databases together with interviews with them while they are alive. The wonderful thing about this project is that the algorithm will fairly make the dead person seem to answer questions when asked now. It's like talking to a person that died several years before. It would be perplexing if this project was realized during the times of Jesus Christ.

Like the 70 years mentioned in the Bible for humans after which illness and other problems would appear, a 2021 Research confirmed that white sharks can live up to 70 years. Cape Cod sharks were used in this research.

Misconceptions must be addressed. People think that politicians, *matais*, academics and government officials have creditable perspectives to guide a country. Not. By adding the perspectives and facts from influencers, activists and the general public (the rightful owners of any nation), meaningful judgements and reckonings can be examined and then include these in the country's goals and resolves as it move forward. The 2022 Court decision that found a former Samoa Prime Minister and some of his other associates

---

[54] Quantum Physics Debunks Materialism. Inspiring Philosophy/ Johanan Raatz channel.

guilty of Contempt of Court, supports my view. Furthermore, the concept of democracy often appears wonderful on paper but usually fail in practice, mainly because it fails to include these types of realities from the masses. A recent example of concepts that appears reputable on paper but doesn't operate well in reality, is the incident regarding four former executives at the Zurich branch of Russia's Gazprombank. It was alleged that these people assisted in laundering funds despite the strict money laundering Swiss laws, known throughout the world banking systems, **as the standards.** Parishioners of every church should contribute to this discussion and maybe advise their legislative representatives.

An interesting case in Washington DC a few years ago where my friend Frederick O'Brien (former US JAG Judge and former American Samoa Attorney General) prosecuted a man who claimed he was told by God to solicit money for Him. During the trial the judge asked the defendant, who gave him the authority to solicit money for God and the defendant said it was God himself and that God is in the Court. The defendant waived to the back of the court for his friend "god" to approach the bench. His friend was a black man dressed in a costume with the word GOD in gold, woven into his attire. The judge immediately shouted to the two to leave his court immediately. At least I have a friend who saw "god". So many sick people are walking around this fine planet that God created and declared "Looks Good" in Genesis, but making fun of the Creator. Some people, especially atheists would comment that this planet is not a **good** place to live since there are now millions of hungry people all around the world. Well, why don't we start by fighting food waste[55], utilize millions of acres of available lands for farming and other developments; mitigate bad politics; jail the greedy people and distribute the wealth. An example of unequal wealth distribution can be seen in Beverly Hills US, where one pair of glasses can cost up to $5k and a watch can cost $10K. This amount

---

55

of money can feed a whole village in some parts of India for a whole month!. Creedy and corrupt world leaders stash away millions of dollars in overseas banks and the taxes from these funds is estimated at be more than 20 billion dollars. These amounts can feed millions of people all around the world. Another classic example is a former Nigerian President who died in 1998, who embezzled nearly 5 billion dollars while millions of people in his country die of starvation. (A documentary about the Pandora Papers during this period revealed that Samoa was also involved in related (setting up corporations without stringent regulations) types of activity and was therefore black listed by the EU.[56]). Many companies perform "sheltering" their profits or as some describe as doing artificial transactions to lower taxes. Billions of dollars salvaged from these schemes can enable governments to fund various social problems for millions of under-served people, not only in the US but around the world. When humans utilize their resources well, then Earth will not be a bad place to live, as created in Genesis.

Sometimes, humans have the answers to these problems but they just refuse to do the right thing and utilize the resources available in an efficient manner to solve these problems. Sharing resources in an efficient global process, will certainly solve most of the problems facing poor countries. There are problems with this suggestion, one might argue. Well, where are the professors and graduate students in the fields of agriculture, economics and commerce? They are around but are thwarted by politics and national capitalistic greed. Do research on these topics and readers will agree that suggested solutions will feed millions or (according to one local farmer) observe the left over from hotels and restaurants, especially after buffets. Contractors and CEOs of some US companies like Lockheed made billions of dollars from the war in Afghanistan. This amount of money can feed thousands of starving people around the world. If one watches the lifestyles of people in Dubai, we may also feel that sharing their resources would

---

56

solve many problems facing countries in the Middle East. I once read a story of how three Samoan men survived in area around Waikiki, Hawaii in the 1960's, mainly consuming cold food obtained from a friend who worked in one of the nearby hotels.

Any human who needs clothes can get some from the Atacama Desert in Chile and from Sao Paulo (readers can research this). Thousands of tons of used clothes from Europe and the US are dumbed in these areas.

Sharing resources, will solve most of the current social problems, and is also a Biblical concept: "For whom much is given, much is required, Luke 12: 48. Selfishness, unfortunately, will always dominate global social wellbeing. It's in some peoples' human nature. Maybe an appropriate form of government for this age is a combination of Communism and Democracy – let's call this *Salanism* for now - where the idea of sharing (from Communism) and the idea of democracy, with its partner capitalism (which has continued to increase the gap between the rich and the poor), should be sampled and execute without the ridiculous caucus phenomenon in presidential elections.

Some people would also complain about droughts all around the world. The technology that moves oil thousands of miles on land can also be used to get water from water-soaked areas to drought areas. Some countries like Namibia, seem to have very limited resources so God gave these people uranium, oil and diamonds. In case one thinks that countries around this area are poor, normally with their barren and desert surroundings and seem to have no resources for their people, just plan a trip to Dubai. (NBC reported during this period that the US is the largest producer of oil in the world). But these solutions would need lots of money some would argue. A chief from my village suggested that maybe the US should use funds from space exploration and arming foreign countries, to care for the homeless living a few hundred yards from their doorsteps. There seems to be no logic in exploring space anticipating mankind's eventual migration to other planets when we need to solve real urgent problems a few yards from one's home and right here on **good** Mother Earth. In the US, there are exclusive affluent golf courses less than 400 yards from areas where scores of homeless people live. Dozens of homeless people, including at least a few army veterans, live in the many tunnels right beneath Las Vegas strip where well-off people spend millions of dollars every night. Furthermore, the sight of several ghettos, slums and reservations all around the US is disgusting (please watch various Documentaries available on the internet). Do we still think we don't have sufficient resources to live a comfortable life in this planet that the Christian God created? or do we just need to adjust democracy and our world view just a bit?.

The religious charismatic movement is also to be blamed for some of the global problems. For example, preacher Kenneth Copeland said on television that God told him (I don't think God ever spoke to him considering his lifestyle and according to Justin Peters Ministries) that his ministry's budget would be $300 million for 2022. These funds can feed all the starving and homeless children around the world. To give some perspective on this amount of funds relative to other ministries, the Grace To You ministry's budget is less than 10% of this amount, yet is has done more wonders in the spiritual lives of people than the Copeland ministry. Copeland may want to consider preaching in the outskirts of countries like Pakistan, Venezuela and India to test his faith, according to a parishioner I interviewed.

Incompetence of humans in managing millions of dollars of public funds also play a significant role in the difficulties faced by millions of indigents around the world. For example, the millions of dollars that fraudulently disappeared from COVID-19 pandemic funds and the millions of dollars in US Medicare fraud cases, can certainly build many homes for the homeless. In American Samoa, hackers stole millions of dollars from the government during this period but no one was fired. Because most of the populations of these islands live below poverty lines[57], these funds would have been

---

57

used to cater for the needs of the indigent locals. I can go on and on with many examples but that would be for another book, maybe. **So, atheists stop blaming the Christian God for humanity's stupidity**, or opt to blame a different god.

There are also many very poor families in Samoa who have occasionally been featured in the local media (e.g., Samoa Observer). It is an uplifting feeling to observe that every time an article about these unfortunate families appears in a local newspaper, various assistances were offered by several local Good Samaritans, businesses and NGOs. This is an essential part that the media must continue to play as a contribution to social progress instead of writing negative "Dear oh Dear" articles that doesn't help the poor. Said entity should consider the more positive phenomenon called "One Small Step" which grew out of a project called Story Corps; an oral history project that started 18 years ago where a half a million ordinary Americans, telling their stories of the need to stop conflicting political views, but come together. This is the largest recording of human voices ever recorded and therefore should be seen as a creditable source of public perspectives. The project brings together different political voices and tries to stop demonizing one another; communicating face to face. "Dear oh Dear" can learn a lot from this project, and from the experts[58].

Why are these issues included in this book?. Because these are all God's people and churches and many NGOs are doing a lot of related work to mitigate these problems but they need the government to forge a new and innovative political and economical way forward. In the US, paying professional athletes and top executives millions of dollars while paying school teachers a meagre $30K is ridiculous. In late 2023, the LA Dodgers signed a $700 million contract with one player! This amount would solve all the problems relating to our underserved population. According to one *Falealili* chief, no human should be paid more than one million dollars as there are hundreds of other people

----

58

who have more knowledge and experience than them, with some yet to be discovered, and that this is not good for the economy as a whole. This is not just a local idea, it is supported by innovative economists like Nick Hanauer, in which he suggested the concept of reciprocity and cooperation to challenge creed and to lessen the gap between the rich and the poor. When the greatest nation on Earth is economically and politically stable, it can do wonders all around the globe.

The main problems on Earth are man-made. Politics, greed, corruption (and sometimes religion in general) are the main culprits. A good US example in regards to politics is allegedly the FDA's ridiculous prolonged vendetta to stop Doctor Burzynski's treatment[59] of cancer patients, where more than 30 million dollars of tax payers' money was spent over several years of congressional hearings and court cases, despite several evidences and testimonies that Burzyskis treatment of cancer actually worked. I would have taken my wife to be treated by the good doctor if I had come across this information before she passed away. The easy solution to this type of problem is to find the source then eliminate it[60]. Genesis 1:31 however, is still correct, Earth was good immediately after creation.

Here are some interesting "*motugāaafa*" – miscellaneous related accounts.

According to some astrophysics, the star that marked the birth of Christ was the result of a "conjunction between Saturn and Jupiter and occurred in 7 BCE in the Pisces constellation.

Jesus was nailed through the wrists not his hands and that recent research suggest that Jesus died around 33 BCE, according to a few scholars.

Some Biblical researchers believe that Goliath suffer from a condition called Giantism (like the famous wrester Andrea the Giant) and because of this condition he had bad vision and that's why he called out to David (1Sam 17:44) to "come to me", so he can clearly

---

59

60

see him. Goliath may also have had problems with his movements in relation to his weight and height so has to be "escorted" down the valley". According to one scientist [61] science confirms Biblical Creation, and he wrote: "It is genuinely surprising that an organism that had evolved by random mutation and selection appears to be designed. It is true, but to many, it will always remain ridiculous". According to recent survey[62] from Britain, two thirds of the people surveyed believe that science and religion are compatible.

According to 60 Minutes Australia 10/1/18, there is a family in Turkey where members walk on all fours, like animals, and have been this way for more than 20 years. Are these people included in the Theory of Evolution?. I thought all of us *homosapiens* have evolved already!. According to one scientist, humans are not the goal of evolution and we are **not chimpanzees but fish**!. Speaking of fish, Samoans knew thousands of years ago, that they may have evolved from fish. According to Samoan legends, and to a local Bible College instructor, different villages have "individual gods" and these are normally some kind of fish. For example, King Malietoa's "individual god" was the fish called "*anae*" – mullet (Dr. Rev. Elia Taase). Dr. Jeffrey Tomkins also noted that 'for about 23% of our genome, we share no immediate genetic ancestry with our closest living relative, the chimpanzee"[63]. Furthermore, he noted that "human and chimp genomes, only 66 to 87% (**not 98-99**%) similar when **omitted data** is factored back in (Dr. Jeffrey Tomkins and Jerry Berman). I personally prefer to be related to Mr. Fish considering our legends. As for the high estimates of 98-99%, I think scientists should find a way to assign these to just the murderers, hopefully.

---

[61] Leslie E. Orgel. The origins of Life, molecules and Natural Selection: (New York: Wiley. 1973) p. 182.

[62] Premier Unbelievable. "John Lennox on science, faith and the evidence for God. 2023.

[63] Ebersberger, I. et. al, 2007. Mapping human genetic ancestry. Journal of Molecular Biology and Evolution.

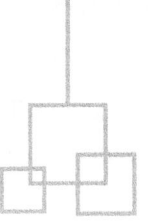

Science and Technology

During this period, I posted a question on Facebook to the two top scientists of this period, Neil deGrasse Tyson and Michio Kaku:

> *"Has man created anything FROM SCRATCH – i.e., not using any "items" like particles, forces, energy etc. that were already present on Earth? the answer to this question will provide an explanation or answer to many questions."*

After more than a year and without getting an answer, someone posted this to me:

*"Sir, its beyond human comprehension to try to explain what was before the Big Bang, I've posted the question to Michio Kaku and Neil Tyson: Has humans invented anything not using any items (subatomic particles, fields, force, energy etc.) that was not already on Earth? I haven't got a reply for nearly a year, **Don't expect one, Good luck".** I know these fine scientists are super-busy and so are some of the professionals in these islands. For example, I emailed the local newspaper Samoa Observer for permission to use some of their materials for this book and never got an answer in more than a year.

Technology had and will continue to contribute to spreading the Gospel of Jesus Christ. During this decade, the concept of "Deep Learning" emerged where computers were programmed to think, assess and improve certain situations. This has a bright future in the field of medicine, Climate Change and maybe religion. Children with genetic deafness can now be treated with gene therapy. Imagine feeding into computer systems all the past and new information and discoveries (some mentioned in this book) relating to the Bible, then allowing Deep Learning do its job. As a former IT supervisor, I predict that in the next 10 years, the outcome from such an exercise will be quite exciting and revealing. On the other side of the coin, the Hubble telescope phenomenon was mainly attributed to Mr. Hubble, but it was the priest/physicist Geroges Memaitre who actually did the related calculations, providing implications of Einstein's general relativity theory that the universe may have begun in a singular point (the initial description of the Big Bang Theory), which cemented the implementation of the Hubble telescope project.

Artificial Intelligence (AI) recently revealed that its programs have enable computer systems to write Children's Books and also make beautiful artwork- some have even won prizes in the Art World!. I'm confident these advancements will benefit humankind and maybe religion in years to come but this innovation **needs to be regulated right now**. Recent use of AI, for example, by entering the passages relating to Jesus calming the storm (Mathew 8), AI showed images (most look like oil paintings) with some showing more than one boat; some showed women in a second boat; and one showed Jesus standing on water. Mimicking artists singing and a US Senator talking about crucial issues can become dangerous and can also lead to fraud. Its cloning attributes is amazing but can also result in serious disinformation.

The most recent advance in technology is the Quantum Computer phenomenon. This system, according to scientists, will have the computing power to solve very complicated problems such as those relating to the Big Bang Theory, in a relatively short period of time compared to current digital computers. The field has evolved from analog to digital and now to quantum, where concepts like super position and entanglements replaces the traditional zeros and ones; the classical building blocks of digital computing. Technology had also helped saved many lives in the past. In 1941, the allies broke the German encrypted codes of military communications and resulted in lessening the period of war by two years, thus saving thousands of lives.

According to Dr. Peter Gay Manners's "Bringing matter to life with sound (1980) presentation, he explained that the effects of cymatic frequency might explain "In the beginning was the Word and the Word was God" according to his thesis 'Everything owes its existence to sound".

Science seems to strip things apart but Christianity explains the "how and why". Pseudo-scientists have been trying hard to disprove the Genesis creation story but they will ultimately fail. Many more discoveries will emerge in the next 100 years that will prove Genesis. Today's science facts maybe proven incorrect in the future and some of these so-called facts and theories keep changing.

For example, several years ago scientists claimed that there was no previous life on the moon as water is required for life to survive. In March 2023, it was announced that scientists have discovered a new and renewable source of water on the moon using lunar samples returned from a Chinese mission. Another example is when scientists recently discovered a giant planet orbiting a massive pair of extremely hot stars, an environment previously thought too inhospitable for a planet to form in. A research article published recently in the science journal Nature noted the discovery of the planet, named "b Centauri (AB)b" or "b Centauri b," disproves a widely held belief among astronomers. "Until now, no planets had been spotted around a star more than three times as massive as the Sun," wrote the European Southern Observatory, which photographed the planet from its Very Large Telescope in the Chilean desert. The study's leader, Markus Janson, a professor of astronomy at Stockholm University, said **"it completely changes the picture about massive stars as planet hosts."**

In 2022, scientists found fossils of early human ancestors in a South African cave that maybe 1 million years older that first estimated – earlier than the famous Lucy or Dinkinsh fossil. Fossils have been one of the basic scientific pieces of evidence used in theories like evolution but recently, some scientists have argued to forget about fossils and concentrate on DNA. Maybe these scientists have found out that floods form fossils fast. Government documents have readily been used as reliable evidence in many court cases. Consider a Canadian project done in the 1960's where children were removed from their families and assigned a different birth certificate, then adopted by various couples. Would these documents be considered as evidence in court?. A similar project done in Ireland called "Irelands Stolen Children fight for Justice" was discussed in the DW Documentary "Forced Adoption and the Catholic Church". All these events have some relation to religion.

Explaining how life on Earth came about involve a few approaches. The two main ones are science and Christianity. These two are sometimes miles apart and therefore would be very difficult to digest in a single setting because of their different basic and stark differences. Oranges and Apples. Science utilizes scientific evidences, proofs and scientific methods while Christianity operates on faith alone. However, in the past decade, a few scientists have advocated that science doesn't' contradict the Bible teachings (several resources available on the web). Interesting though, science has in some instances proven some components of the Biblical scriptures. For example, the existence of people and places mentioned in the Bible, the Census mentioned in Luke; the slaughter of babies by Herod[64] and the science behind the benefits of fasting. (Also remember philosopher Socrates said he fasted so he would get clarity for his mind). According to the Biblical Archaeology Society, archeologists in northern Israel had discovered the stone slab known as the Tel Dan Stele, in 1993. The inscription on the slab is evidence of the

---

[64] Dr. Sean McDowell, Dr. Kennedy Titus. :The Archeological Evidence for Jesus". 2021.

existence of a royal house in the Ancient Near East that bore the Biblical character's name "David". Dating of burned fragments (around 1410 BC-give or take 40 years) revealed that the ancient city of Jericho did experienced a destruction event around the time indicated in the Bible, according to a report by the New York Times. Another recent discovery explained the existence of Nineveh, a place located on the River Tigris. Detail description of this discovery can be found in the book "Where God Came Down, by Joel B. Kramer.

As students all over the world continue to study evolution (they should continue to do this, then decide for themselves), they should be aware that Darwin didn't know much about DNA. (Dr. Sean Caroll stated that Einstein didn't know much about black holes). Students studying science (including Samoan students) now know more about DNA than Darwin!. He didn't know the principles that "you can't select what's not there" and that "natural selection is not evolution". I'm not surprised since Darwin's degree was in Theology and according to Dr. Jennifer Hall Riviera, Darwin was not a scientist and was not trained in human anatomy, and if there are still people out there who continue to believe in Darwin's Theory of Evolution, please continue reading this book and also examine Dr. Meyer's book "Darwin's Doubt" and Dr. Jason Lisle presentation "Science Confirms Biblical Creation" which provides additional related arguments against the theory of evolution. Additionally, readers maybe interested in Dr. Jeff Tomkins presentation "Debunking Evolution and Proving Creation". The many articles, peer-reviewed papers and some recent discoveries don't seem to make top headlines these days mostly because the **tenured professors don't want to lose their jobs.** For example, some researchers are not familiar with a study published in 2018 "Why should mitochondria define species?" (Stoeckle M.Y. Thaler D.S.), its the **largest DNA study ever done (the scientific community should all be aware of this),** which suggested that there was no inter-genetic relationship amongst species as one would expect from Darwin's model and that all animal and human life arose around the same time. This was published in the Journal of evolution.

Darwinism doesn't explain the existence of life, and life can't just evolve from nothing. There has to be something to evolve! Dr. Donald Hoffman said that he didn't believe in most scientific theories, and as for Darwin's theory, people can just take it to be true **because it's the best one we have**. For a more stark remark, Dr. David Berlinski said Darwin's theory is all wrong.

The Bible, however, mentions several scientific facts. For example:

i. The Earth floats in space – refer to Job 26: 7, "…. "He suspends the Earth over nothing".

ii. Isolation during the COVID-19 pandemic stressed the importance of isolating those affected. Consider Leviticus 13:1-5 "to isolate the affected person for seven days and if they are not better, "to isolate them for another seven days". It was astonishing that some of the Samoan pastors refused to obey this Biblical recommendation. From my observation, these are pastors who didn't attend Piula, Kanana Fou or Malua Theological Colleges.

iii. The universe is made of invisible particles – refer to Hebrew 11:3 "By faith we understand that the universe was formed at God's command, so that what is seen was not made out of what was visible (consider neutrons, electrons, protons, fields, energy, subatomic particles etc. – invisible to the naked eye - that make up pretty much everything).

iv. Water changes state – refer to Job 36:27: "He draws up the drops of water, which distill as rain to the streams….". Psalm 135:7 "He makes clouds rise from the ends of the earth; he sends lightning with the rain and brings out the wind from his storehouses".

v. The core of planet Earth is hot – refer to Job 28:5: "The Earth, from which food comes, is transformed below as by fire".

vi. Planet Earth and celestial bodies will not last forever – refer to Mathew 24:35 "Heaven and earth will pass away, but my words will never pass away".

vii. There are mountains underwater (Mount Mauna Kea in Hawaii maybe the tallest) – refer to Psalms 104: 6 "You covered it with watery depths as with a garment; the waters stood above the mountains".

viii. Not all the stars are created equal – refer to Corinthians 15:41 "The sun has one kind of splendor, the moon another and the stars another, and stars differ from star in splendor".

ix. There are countless stars in the sky – refer to Jeremiah 33:22 "I will make the descendants of David my servant and the Levites who minister before me as countless as the stars in the sky and as measureless as the sand on the seashore".

The Bible doesn't have all the relevant stories about Jesus Christ, because there were no comprehensive and formal written recordings of ALL related events during that period. Like the Samoans, the Israelites have many of their traditional stories "stored" in their oral story-telling legends. There are also other sources of Biblical-related stories[65]. For example, the Old Extracanonical texts like the "Infancy Gospel of Thomas", which tells a story about the life of young Jesus *(Jesus, Mary, and Joseph: Family trouble in the infancy Gospels)*: The Conversation where Jesus healed his brother James from a snake bite. The text also mentioned a family accusing Jesus of pushing one of his young friends named Zeno, from a rooftop where they were playing, but Jesus later brought back Zeno to life. There is also the controversial Book of Enoch.

More than ten years ago during one of my presentations for a US federal National and Oceanic Administration (NOAA) meeting, I told a Hawaiian presenter that the behaviors and cycles of a certain species of fish that she included in her presentation have been studied

---

65

by Hawaiian fishery biologists and what she described as a Hawaiian legend is actually scientific fact.

Consider the advances in Artificial Intelligent(AI) where the hosts/robots do exactly what the programming code instructs them to behave but sometimes a mechanical error (sometimes due to the programming code) forces the robot to function in a way that was not intended. But sometimes a code in another part of the main program would interfere. AI also resulted in misinformation during this period. Around March 2023, images of Pope Francis produced using AI, appeared online, wearing trendy outerwear. This phenomenon should be regulated **as soon as possible** as it has negative impacts on communities including churches. In fact, the "godfather" of AI, Geoffrey Hinton, and many AI experts have signaled serious concerns about the rapid initial use of AI by "bad actors using it for bad things", and the guaranteed threat to society. This phenomenon has also recently ventured into the US legal system where two Manhattan lawyers blamed the ChatGPT for tricking them into providing fictitious legal research in a court filing. A robot EveR 6, was also able to conduct an orchestra in Seoul Korea during this period. Imagine pastors getting their sermons from AI!. In fact, I have already produced a sample sermon on the creation from Genesis using ChatGPT AI which I will present at the launching of this book.

For members of the clergy who are not very familiar with AI, it is a system that uses mathematical algorithms (set of step-by-step instructions embedded in computer software) that sort, filter and select from a very large data base. It is a machine learning system that takes in information about the past and makes decisions or predictions. In the field of medicine, it can now make better medical decisions than the best doctors because it allegedly utilizes thousands of x-ray images in its decisions compared to few used by ordinary doctors. On a negative perspective, it can increase social controls on the public like that in China. According to one AI expert "Cognition means thinking; your machine in not thinking and AI means a lot of brute force computation". A 3-year-old child recently played Catherine Rollin's Love Theme on the piano and accompanied by an orchestra. **A robot can't do this without the algorithms and electricity from humans**. A young cheerleader name Sarah woke up on her own from a 20-year coma. Robots can't do this as evidenced by the need for a new battery to start a vehicle that has not been used for 20 years, let alone the refusal of the engine to start turning on its own.

During this decade a young man posted a video showing a vehicle that runs only on solar energy and without a motor. Interesting though, the vehicle travelled faster than the wind pushing it! A UCLA physics professor claimed that this was impossible so the two made a bet of $10k that if the man can prove this phenomenon, the professor would give him $10k. The young student invited two of the world's top scientists, Neil DeGrasse Tyson and the Science Guy to be the referees of this experiment. The young student won and was awarded the $10k. Apparently, the physics professor didn't fully understand basic physics equations and concepts. Readers can also do research on this event on the internet. This shows that many atheist scientists (not the theists scientists) don't understand what they're talking about and should stop debating the Christians until they **fully** understand all the laws of physics and laws of nature, all of which God manufactured. Samoans students aiming to study physics at overseas universities should just dump the UCLA Physics Department. I however, understand the situation with the professor from the atheists' point of view that "science is in progress", and that atheists have nothing definitive at this point in time. Fortunately, there are answers in Genesis according to some theist-scientists named in this book.

As philosophy start creeping into technology, new perspectives are being formed in the field of computing, where there are now ethical questions that AI need to entertain. Actually, its up to the algorithms - that will be done by humans - that will provide the basis for some lifeor death decisions. For example, the increasing use of driver-less cars demands that decisions need to be made when a child runs in front of these cars. The car can be programmed to swerve or hit the brakes hard. Both decisions can potentially lead to death. So now, technology is given the authority to decide who to kill. Alarming, yet it's now reality. The person the AI decides to kill maybe a pastor!, but the initial basic algorithm that started the decision was made by humans.

Some laws of physics referred to as very strong theory that will stand the test of time, seems to be against other scientific assumptions relating to the creation of the universe. Readers can refer to Dr. David Menton's presentation "Evolution, not a Chance" on the Web. It was pointed out that the statement "the typical mutation is very mild, it usually has no effect, but shows up as a small decrease in fertility or viability". Well, it should increase according to the law of thermodynamics – contradiction my friend. Speaking of thermodynamics, Spike Psarris noted that thermodynamics confirms creation.

Most of the theories like the Chemical Theory, Stellar and Planetary Evolution; Organic evolution and Fine Tuning[66] are some of the theories that all have deficiencies (some very serious). Fortunately for the Christians, all these are explained in Genesis 1. Christians don't need scientific proof so WE'RE GOOD. One brilliant attorney[67] told me that, in court, lawyers should only ask questions that they know the answer to. Atheists should ponder this concept when debating Christians.

When these scientists said that there are only two sexes, I was confused. When a chief of my village holds presentations for the youths near his taro plantations, he sometimes mentions fresh water protozoa called _etrahymena_, which are oval-shaped microscopic organisms. These have seven different "sexes" or mating types. Some species of fish like the _kobudai_ can switch sex permanently at a specific

---

66

67

point in their lives. The **_Auanema sp_**. has three sexes – female, male and hermaphrodite. There are also humans called intersex and according to experts, 0.05 to 1.7% of the population are born with intersex traits. These are people that have external or internal sexual organs that are clearly male or female.

(A little humor for this section: Someone asked a friend if he/she was a male or female. The friend answered that he belonged to a group that loved girls but girls don't love him back. Maybe he belongs to the sex of rejection or lonely). Someone might argue that these are not humans, but I thought they believe these organisms will eventually evolve into humans?. I can go on and on about this topic but this is not a science book so I'm inviting atheist scientists to these free lectures that take place in the village of _Salani_ every six-months, before the ailing Dr. Lolo dies.

I know Dawkins can't even explain the growling and levitations (one of our AOG pastor personally saw this) during exorcism, let alone the parthenogen phenomenon which certain creatures like the king cobra and sawfish have been known to accomplish "virgin births".

Sometimes the Christian God utilizes or manipulates nature to perform miracles. Jesus walking on water is not a miracle _per se_. Jesus was confident he could walk on water because He created water. (Theres a species of lizards that can walk on water and torpedo fish have their own electricity).

In the last three decades scientists have uncovered clues regarding the plagues in Egypt, mentioned in the Bible. Additionally, scientific explanation of the sea turning red[68] in Exodus may have been due to underground gas leaks, according to some archeologists. This "color" phenomenon should be very intriguing to the people of the Manu'a islands. They have a traditional saying or proverbial expression referring to their traditional king (_Tuimanu'a_) shoreline sea tide as "having a natural yellow color". I believe that this is

---

68

not just a legend but a scientific fact. Native Sulphur is a **yellow** crystalline solid that has historically referred to as brimstone and this substance can be found at volcanic fumaroles where H_2S and SO_2 gases are emitted. Volcanic activities at the Manu'a islands in the past may have caused Sulphur output which is **yellow** and could have prompted the Manu'a orators (*tulafale*) to formulate said adage. Furthermore, the gas Sulphur hexafluoride, is one of the longest-lasting climate pollutants that can remain in the atmosphere for thousands of years. Serious volcanic activities in 2022, that brought down US federal scientists to monitor the associated quakes around American Samoa, should have strike a chord regarding this phenomenon and jog some memories regarding the origin of the traditional axiom: "*tai samasama o le Tuimanu'a*".

Scientist Neil deGrasse Tyson believe that science has solved many basic questions regarding nature, (let me mention tsunamis and earthquakes), then he said that, watching hundreds of people being killed by earthquakes and tsunamis makes him believe that God is not good. I thought they already scientifically solved and explain how tsunamis and earthquakes happen. Now they turn around and said its God's fault. Confused people.

The teaching of evolution should continue in public education. How it is viewed should be left to the students, not to the old professors. The main impediment to the teaching of this theory is the difficulty in perceiving a phenomenon that took millions of years to evolve, while students nowadays need evidence and data they can see and grasp during the lesson. Keeping one grounded in observation is always good for science students.

After all my research on science as it relates to life and in regards to explaining the origin of life on Earth applying the latest discoveries, I've come to the following conclusions:

Physics, maybe the queen of science, *is moving very slowly forward as there are still many questions regarding the Big Bang Theory and there is also the question of 'what is energy?'. A group of top*

Theoretical Physicists agreed that we still don't know everything about Black Holes but we're getting there. Maybe in the year 3000 I suppose.

Biology *has come to a near standstill after several serious questions regarding the Theory of Evolution. For example, see Dr. David Menton's presentation: "Evolution, not a Chance", where he also mentioned the top Genetic Scientist Dr. James Crow's view regarding evolution. Dr. Menton also mentioned that atheist Richard Hawkins seem to contradict his convictions when he said that "Biology is the study of complicated things that give the appearance of having being **designed for a purpose**. This again proved my view (and also Dr. Menton's view) that this atheist "seem to be a person who knows more things that aren't true than any human being"* (Dr. David Menton. 1997).

Genetics however, maintains that we are all Africans!.

Chemistry *has, in general, concluded that there must be an intelligent design that created the heavens and Earth (and also according to Professor John Lennox, Professor James Tour, Dr. Denis Lamoureuz, Dr. Andrew A. Snelling and many others).* **Astrophysics and Quantum Mechanics** *is asking scientists to observe the human consciousness and astrophysicists have developed their own "Periodic Table" that include items like neutrino, up quark, down quark and others, and are warning mankind of it ruling the world when it teams up with Artificial Intelligence in the very near future. I feel sorry for Chemistry and its Periodic Table.....[while a number of philosophical ideas] maybe logically consistent with present quantum mechanics,..... materialism is not"* (Physicist Eugene Wigner). **Mathematics,** *the purest form of knowledge, is just smiling at the debates and saying "the beauty of infinitive complexity is built into numbers and that the secular worldview cannot account for the existence and properties of numbers or mathematical truths".*(Dr. John Lisle. Indian Hill Comm C. 2019). *moreover, the miracle of the appropriateness of the language of mathematics for the formulation of laws of physics is a wonderful gift where we neither understand nor deserve".* (The Unreasonable Effectiveness of Mathematics in the Natural Sciences. Dr. Eugene W). *The contention of most of these people is that mathematics is the "queen"* *of science and can explain life better than other sciences and its Langlands theory of bridging the Number Theory and the Harmonic Analysis may one day bring some light into the mathematics involved in Genesis. They're still pondering on the 1/137 "fine structure Constant" and are happy that their field takes one" beyond the capacity of humans senses". According to Professor David Gelernter, there is still a mathematical problem with Darwin's Theory of Evolution and that there is too much inferences. How about the P=NP computational complexity problem and a continuation of Alan Turing's 1952 equations?. Scientists need to resolve these issues before participating in any debate with Christians.*

__In medicine__, Neurosurgeon Eben Alexander III MD cautioned on what he called "reductive materialism" utilized by many atheist-scientists in their arguments. Many of them have also warned on national television, of voodoo-science being used to argue against Christianity.

__Engineering__: Human skeletal joints are masterpieces of engineering. Dr. Stuart Burgess of Bristol University, UK who has studied __biomechanics for 30 years__ stated that humans skeletal joints are __design masterpieces__ and only go wrong when they experience a disease like arthritis or they have been mis-used or overused, but __not because of design__ and looking at the elderlies of 70 plus years (like this author), one should be in awe of the Creator's masterpieces in engineering. If anyone thinks human skeletal joints just evolve into engineering masterpieces, they're complete idiots. Ask Dr. Burgess.

The Christian __God doesn't need defending though__. Jesus himself also refused to allow Peter to physically defend him, telling Peter to put away his sword. HE also doesn't need defending why bad things happen on Earth he created since he created this planet and saw that it was "very good" Genesis 1: 31- consequently humans should ask themselves, not God, these questions.

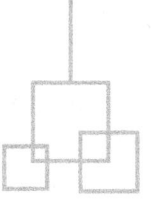

Illcit Drugs

I llicit drugs include marijuana, cocaine, heroin, amphetamines and several prescription drugs. This topic is included in this book since it is the opinion of this author, and many pastors, that the churches should increase its involvement and role in programs related to the national efforts to curb this threat to our local community. In 2022-2023, drug related matters were mentioned in local news nearly every day. Respective villagers and their Council of Chiefs had played a major role in these efforts in the past decade but more has to be done to safeguard the community. Unfortunately, politics and corruption at the highest levels of governments in these islands has hindered related progress. If a Governor or Prime Minister offers a friend of mine the position of Police Commissioner, he will definitely clean-house in less than a year, but would probably resign before then, because there will be so many difficulties emanating from top government officials and law enforcement employees who are drug addicts themselves.

A few years into this period, Samoa saw the war on drugs in Upolu reached its peak when former Commissioner Fuiavaili'ili Egon Keil, was the Police Commissioner. The problem suddenly declined significantly as this experienced Police Commissioner went straight to the root of the problem and fired several high-ranking police officers. Additionally, a late 2019 raid involving 100 plainclothes police officers, busted a drug ring in Faleatiu, in what was billed as the largest drug bust in the country's history. Police, guided by aerial drones, swooped to seize some 10,000 marijuana plants, firearms, methamphetamine and an undisclosed amount of currency. An armed standoff was eventually defused before six people were taken into custody. About a year later, similar scenes played out when police confiscated 4,000 marijuana plants and three suspects following another armed raid in the mountains of Satapuaala and Faleatiu. These were nobly motivated and just attempts to kill off the nation's problem with illicit drugs at its source.

Samoa then learned about related phenomenon similar to those which American authorities have learned over the last half-century as they led a global war on drugs. That phenomenon is something that economists call displacement. When demand for a product, legal or illegal, is high enough, and the rewards of its production are too, any clamp down on its production in one area will simply result in an increase in another. It is now being alleged that one of the most senior bureaucrats in the islands was using drugs and doing so indiscreetly. Raids in one country would simply push production to another; crackdowns in one neighborhood would push drug sales to another. Since the reported enormous hauls by law enforcement, Samoa saw a steady and very similar examples that are proof of this principle in action. Drugs did not stop entering Samoa despite the regular drumbeat of news about the arrest of those involved in their consumption or distribution. A few years later, a joint operation by the Samoa Ministry of Customs and Revenue and the police foiled an attempt to conceal methamphetamine consignments inside consumer packages on shipments coming into Samoa from overseas. Despite the size of the seizure, it apparently did little to stop the drug's presence on the streets. A few weeks later, a raid on a safehouse resulted in the arrests of suspects allegedly involved in a conspiracy to distribute it on the streets. Another $20,000 worth of methamphetamine was seized after police raided the same Ma'ali Street property for the second time in a year. The drug problem has since escalated and will never be curtailed, affecting many parishioners in these islands.

In early 2022, the services of a few police officers were terminated in American Samoa when they tested positive for the drug "ice". This pointed operation was directed by Commissioner Chief Lefiti and not under the leadership of the previous younger Commissioners with relatively higher education, but had refused to follow-up on the various related cases (after many years in office) and were strongly influenced by bad politics. Drug problems doesn't require highly educated leaders, it just needs ordinary people with elevated ethics and extreme integrity. Several months later several American Samoa Government (ASG) employees tested positive for ice. This problem may gradually be solved now that the investigations are approaching the "sources". Several police officers in Samoa were terminated around this time due to the same problem. A good CCCAS pastor didn't mince words during his KVZK Sunday television service around this period when he said that the leaders of the country are involved with drugs and that was the main problem.

In the morning of October 13, 2023, a pastor of the Potters Church of Iliili gave a strong testimony on the local KHJ radio station regarding drug rehabilitation. He told of the few related programs that he participated in while living in the US including the famous Betty Ford clinic. According to the Pastor, all these programs didn't work for him. It was during a church service he attended and was saved by Jesus that enabled him to stop using drugs.

During this period, the presence and use of heavy drugs in Samoa and American Samoa skyrocketed and neighboring Tonga was also being ravaged by the drug trade and become one of that country's foremost social and political issues.

Again, the churches must step in to play its role in solving this public problem. Personally, I think this is the most substantial predicament in these islands.

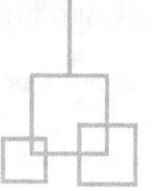

The Confused Atheists

The traditional Samoan family unit, under the chief or the man of the house, traditionally dictates which church family members join. The church has a strong influence in the lives of the people of these islands and one wonders if there are atheists within the local community with its two governments declaring the Christian God in its formal national slogan. My experience and scores of contacts and conversations with several people for several years has confirmed that there are a few atheists amongst us. I've known a couple of these people for more than 10 years but they sometimes attend church because their parents demanded that they do. My many conversations with these well-educated locals clearly reflected their atheist views and have continued to this day. Maybe there are more out there and have yet to "come out". This section of the book is very important to me because it reflects the concept of *agabe*. It seems that atheists continue to preach their views and blaspheme all over the world while at the same time living relatively comfortable lives compared to the indigent Christians who struggle in life trying to survive poverty and oppression. God should eliminate these people immediately, one street beggar told me. I disagree. Apparently, the Chrisitan God even considers these lost souls and has inclusive love for atheists. A concept even I can't fathom.

Let me give the reader my definition of atheist as gathered from several years of research. Atheists are people who don't believe in the Christian God and are characterized by their mechanical, computer-like (garbage in -garbage out) personalities that lack a soul and refuse to "think outside the box". They refuse to wander outside science, beyond the Big Bang. The box being the sphere in between the Big Bang and afterlife (refer to my Cosmic Chart in another Section). Because they argue against scientists in televised debates, they therefore have no solid grounds for their arguments because their ideas are mainly based on science examinations **in progress**, with mostly just theories. They use mainly science for their arguments yet disagree with top scientists during debates – very confused people. These people generally lack joy (they have no songs like Gospel hymns), hopes and optimism as they argue against people with established convictions whereas they have nothing (many, like Dan Barker, declared that they **lack** faith) worthwhile to contribute to the general peace and harmony of society, but instead live **to attack Christianity as their only mission in life**. They live to destroy and attack Christianity but offer nothing constructive for world peace and coherence. After investigating and following their general demeanor and attitudes for a few years, I feel sorry for these "dudes" and conclude that they appear to be very sorrowful, arrogant and depressing humans. Readers are welcomed to do their own research and conclusions. I'm still wondering why the Christian God continue to allow such blaspheme attitudes to continue. Jesus knew about Judas before he selected him as a disciple. Maybe his divine plans include "an opposite and equal reaction". Maybe evil is essential and that if there's no God there will be no evil. If there is no light then there will be no shadow. If one sees "not-so good" people in church, it's OK. The church should be a hospital for sinners. For those scientists (especially those in the field of chemistry) who still doubt the "intelligent design" concept, please visit Dr. James Tour for some guidance.

The following chart may illustrate the various areas of beliefs and related practices which should help readers understand the different perspectives related to religion in general:

The *Malaloa* Analysis

Following is my simple chart of religion-related topics and major players who have contributed to related televised discussions.

THEISTS - e.g Professor John Knox, Dr. Stephen Myers, Dr. Jason Lyles
GNOSTIC INFORMANT, REALIST, HUMANIST - e.g Dr. Richard C. Miller
ATHEIST – e.g. Richard Dawkins, Christopher Hitchens, Dan Barker
WITCHCRAFT, MINDCONTROL, DEMONIC PRACTICE (Watch a congregation run out of the church and start eating grass (IsaiahSaldvar. Prof Lesogo Daniel. Rabbon Canter Ministries)

I have an interesting Samoan illustration that I want to share for the sake of atheists (hope some will read my book) who make fun of Christian beliefs. Many non-Christians often wondered why Christians have faith in something or entity they can't see. Some are amazed as to why Christians believe in a concept that has no scientific proof (there are scores of scientific proofs available from several sources, just do research on the web). Apparently, these atheists do the same thing or behave in a very similar way as Christians. For example, imagine a Samoa Airways plane flying over these beautiful Samoan islands. The pilot maybe the only one who understand that the difference in air pressure at the wings provides the "lift" necessary to lift up the plane while the engine just provide the "thrust" or push needed to move the plane forward. An aviator may find the quote in Exodus 9:14 interesting: "You have seen what I did to the Egyptians, and how I carried you on eagles' wings and brought you to myself". This metaphor can be interpreted as the powerful and quick deliverance by God of his people. The pilot can't see the "pressure" but believes that it exists and is providing the required "lift". The is similar to the way Christians believe in their God. They can't see Him, but can experience His effects and impacts. Most passengers on this plane have no idea about pressure – these are the non-believers in the Christians God, or those that just don't care, but the "pressure" still provides a safe trip for them - God loves everyone, and may you have a safe trip.

In my previous book, I documented a short email debate with one of the most well-known atheists in the world, Dan Barker. It is very uncommon for a person from these isolated pacific islands to dispute a distinguished expert in any field, let alone the field of theology. However, I personally feel compel to engage this poor soul so he might see the light, from a Samoan Chief's perspective. Barker should be seen as a human **much more intelligent than the Christian God** because he said on national television that "there are some serious issues with the Ten Commandments". This is one of the reasons I debated this atheist, because he is full of dark convictions. He also said that Eve (in Genesis) was framed. Since he depends on scientific evidence for his arguments, where is his scientific evidence that Eve was framed? Now I agree with one of my friends that this man is nuts. He also maintained that Christians have the burden of proof. Not according to my chiefs. They said that the atheists have the burden of proof. Additionally, Dr. Alister McGrath agreed that Dawkins and other atheist should have the burden of proof when participating in debates, and that atheism is not privileged. It is a system of belief and therefore need proofs.

Dan Barker (or any other person) makes their own rules. Christians utilize the rules in the scriptures. We make our own rules in our village. There is also a challenge that Barker should be interested in. It states that: "If the Shroud of Turin is a forgery, show how it's done" (this shroud is the most-studied archeological object in the history of the world, according to Father Andrew Dalton). I know Dan Barker, can rise up to the challenge and get the $1 million reward. I also have a Challenge: If any atheist can create a planet and placed it 5 inches from a Black Hole, along the "event horizon", I will forsake my Christian conviction immediately. I offer no monetary reward though, since I'm a poor Samoan Chief. One of my friends even offered this personal challenge. If atheists can produce a single living cell in a laboratory he will stop going to church. When I asked him who created God he looked at me and said, well, who created the "cosmic soup /egg" that many scientists believe started their whole creation theories?.

Atheist debaters are usually distasteful humans who mainly don't provide personal evidences to prove their positions and many of them declare that they lack any kind of faith. Their only mission in these debates is arguing against theists but don't usually declare any **personal** solid grounds regarding their positions. Most Christian debaters provide evidences, testimonies and related discoveries while atheists debaters concentrate on attacking theists' arguments. Here's a funny debate between Professor John Knox and Professor Peter Singer.

> **Knox:** my parents were Christians ….
> *Peter:* There goes my main objection against all religion, people stay in the religion in which they were brought up.
> **Knox:** were your parents atheists?
> *Peter:* yes
> **Knox:** So you stayed in the faith you grew up in?
> *Peter:* But it isn't a faith.
> **Knox:** I'm sorry, **I thought you believed it**
> Peter, an atheist, is confused.

If readers think that I'm too hard on the poor atheists; consider the following quote from one of the greatest atheist in the world, from his book "The God Illusion".

"The God of the Old Testament is arguably the most unpleasant character in all fiction: jealous and proud of it; a petty, unjust, unforgiving control freak; a vindictive, bloodthirsty ethnic cleanser; a misogynistic, homophobic, racist, infanticidal, genocidal, filicidal

pietistical, megalomaniac, sadomasochist, bully...... ". Therefore, I also have the right to call any atheist, any name. Fair enough?. I hope the publisher would also be fair and leave this text for the readers to decide and for their assessments. In fact I've toned down several comments in this book, just to get it published.[69]. This book is full of inaccuracies and Dr. Adam Rutherford seem to agree with my analysis.

There are authors who wrote several books and then renounce them later when they get old and wiser. For example, Philosopher Herbert Fingerette wrote a book about Death and later when he was facing death at 97, he rejected many of his perspectives on Death. I don't think Richard Hawkins will ever change his mind because he's one of the growing number of atheists that even if evidence contrary to their convictions are provided, they will not budge. Conflicting and stubborn personalities. At least Christopher Hitchens stood his sludge - I mean ground.

I had also watched many debates between atheists and Christians on television and find these exchanges very stimulating but sometimes irrational. One such debate was televised from the Berkley Center where the famous atheist Christopher Hitchens made some very silly statements. He stated that religion was the cause of many wars and injustice acts in the past, including World Wars. Well Chris (sorry I have a young nephew name Chris) Hitler may have been a Catholic but **human nature** and **the free will**- that was in his DNA was the **root cause** of all the demonic things he did, and absolutely NOT Christianity (maybe other religions). As for the notion that he was a Catholic, he was definitely not a devout one. There is no Biblical demand in the Gospels to murder millions of Jews including thousands of children. Wars thought to be caused by Christianity is absolutely ridiculous. In these modern times, the Christian God didn't call for wars and killings and there is no text in the Bible mandating these wicked acts. Additionally, not every

---

69

catholic parishioner prayed for Hitler, as Chris mentioned. People of other religions (he mentioned there are about ten thousand different religions in the world) didn't pray for Hitler and **definitely not the people of Samoa**. According to the Encyclopedia of Wars, up till around 2002, there were 1,763 wars and of those, only 8% were caused by religion, therefore 92% were secular wars. Again, the reference to religion shouldn't include Christianity because the leaders of these acts' mis-interpreted the Bible, and therefore it's the leaders' personal conviction not the concept of Christianity that allegedly cause wars. Some examples on wars that are commonly referred to as "wars in the name of religion" include the Crusades and the Spanish Inquisition. Related events include the slaughter of Christians in Roman coliseums, Catholics killing of a certain group in France, burning of witches in Massachusetts, the prolonged killings of people in Northern Ireland during the conflict between Protestants and Catholics, Indians and Muslims killing each other in Pakistan and the continuous conflicts between Jews and Arabs. The Christian God had nothing to do with these events, it was the free will of the brats - sorry I mean bad humans.

There are however religions that call for wars and killing. One can watch podcasts by Dr. Jay Smith to get an idea of this fact. The main fault with these television debates is the Christians' squander in trying to explain what God meant or why he did certain things. My dear Christian debaters, please stop this nonsense as you or anybody else can't read the mind of the Christian God. For the atheists debaters, you have absolutely no glue as how you should interpret the events in the Bible and also can't prove that the Christian God **doesn't** exist!.

When Chris Hitchens said that humans have nearly the same genomes as other animals, I quickly remember a conversation a young American Samoa Community College (ASCC) student had with his engineering instructor when the engineering class visited Solo's Jr. autoshop where scores of cars were parked all over the compound. The student asked the instructor if the three very similar cars near the gate were made and designed by the same company. The instructor said yes, they were made by the same company but in three different countries. Furthermore, the vehicles though they all have the same basic engine constituents and design, one of the vehicles was a 4WD, designed for rough terrain and one has automatic transmission and GIS technology, the other with future models that will be driverless. Different purposes and capabilities but look very similar in basic design.

Chris normally underscores his approach as having **doubts** towards any Christian contention. This "doubt" demeanor reminded me of various discoveries of several ships found in the **middle of deserts** in countries like the US and Africa. Chris would doubt ships will normally be found in the desert, but then there it was, clear photos and well related documentations. Some of these rotting ships still contained some treasures when these were discovered in the desert, of all places!. Very strange but factual and well documented. Chris wouldn't believe that there were sub-atomic particles (before they were discovered) because he can't see them, but these undoubtedly were there all along. When one finds a great piece of art, the item can't speak about itself or for the artist, but it was definitely created by a master. An explanation of any great item has to come from "outside the object". The explanation of the Christian God must come from "outside the box" and no object on earth can be explained from within any object. The undeveloped brains of these people always find Biblical concepts bizarre and I understand that perfectly. Some scientists believe the concept of cloning animals has resulted in the creation of new animals. This is absurd, cloning is just transferring information, not creating new species and has always resulted in weird animals like the famous Dolly. The famous professor is welcomed to debate me under the huge coconut tree near my *fale*.

Chris also accused the Muslims of some of the unfortunate activities that happened to some of his friends stemming from the Muslim religion writings. It should also be noted that many unfortunate and very evil events have also occurred to many people because of the atheists (like Chris) convictions (please do research on the web). Furthermore, Chris stated that people who believe in creationism believe in it because it makes them **feel good**. Chris, I believe in the God of Abraham and his creation but I don't feel good about it sometimes because of various realities that I will discuss if you want to debate me. But just to mention one reason. The **consequent** human nature and the implanted free will, after creation, may eventually result in the presence of a human monster who may kill the people of my village. So, I'm not really thrilled or **feel good** about intelligent creationism, but I, however, believe in Jesus Christ.

Chris also "measures" a church's position by the statements made by its religious leaders and he gave some examples of what the Catholic church leaders said. Chris, sometimes Church leaders err. It's human (you err so many times on national television and I can point those out) and many scientists sometimes change their views. Sometimes, the Christian debater misses some crucial point during debates but their truth is ultimately with God. Quoting **just one** Catholic parishioner maintaining that the Northern Yorkshire flood was a punishment for homosexuality is childish and silly. Even one of my orators (*tulafale*) from the village of *Salani* who seldom goes to church and didn't finish High School found Chris's statements ridiculous. Atheists give varying statements and perspectives and I can't judge the whole group of intelligent and fine atheists by one silly and stupid atheist's statement.

This atheist (and many others) also stated that humans can all be moral beings without religion. Writing for Christianity only (I'm not an expert in other religions), he failed to include the reality that hundreds of "undecided" people have become moral beings because of Christianity; hundreds more who were tempted to do evil acts refrained; and hundreds more learned to forgive because of Christian teachings (readers are encouraged to do research on the internet on this subject). Mybe Chris was referring to people with different genre of morals that caused the 9/11 catastrophe?. Readers should also note that Dr. David Berlinski once said that Chris "was wrong in a lot of things".

My comments above may seem offending but then the debates mentioned above were done on US national television and viewed by millions of people where atheists blaspheme against the Christian God. The most important facet of these debates is my observation that Chris was an arrogant human being who believes he has the answer to every topic of the discussion and would often make fun of the opposing arguments. I also have this right. His conviction that reason and scientific evidence is the ultimate "yardstick" on which every argument must be based on, is his own personal conviction. Absolutely not mine and millions of other people like some science professors, who have their own yardsticks or paradigms. I have my own and would happily discuss this after an *ava* ceremony to welcome him if he chooses to visit these beautiful islands. I want to underscore in this book that people of all colors and background have different beliefs and views but that the important challenge is to respect each other's perspectives and convictions since we all live in this relatively **small** planet (compared to the universe) and therefore harmony is essential for our existence. By **small** I refer to the 2023 discovery by scientists that "each of the six objects they recently discovered through the James Webb telescope weigh billions of times more than our sun". We are "small potatoes" according to my friend Palela Pule. Religion and the rule of law helps to bring harmony and order but it's more efficient to rely on respect, kindness and a social system like our Samoan culture, to make this world a better place.

All of Christopher Hitchens (also Richard Dawkins and all other atheists) questions about the Christian God can be answered by this one personal statement<u>: The Christian **God does not follow reason and science, and he can do anything HE pleases because HE is God.**</u> Any human that has thousands of questions that can be answered by one statement is a questionable character. This is in contrast with a video presentation on the internet called "Why

Christopher is great". Another notable old atheist-scientist[70] said that even if he is given scientific proof of the existence of the Christian God, he will still not believe in God. Since he, as a scientist, require scientific proof, then his statement mirrors a very disturbed human being. Additionally, he stated that "science deals with questions where there is evidence". Well, why do some scientists need to ask questions when there is already evidence according to them?. I recommend that atheists read the books "Ireversible Damage" and "Material Girls" by Katherine Stock to educate them in regards to their statements about humans having only two sexes.

A comment by a Martin R. regarding so-called professor[71] F. (another notable atheist) is that that this atheist is a funny and smug arrogant guy who is like the Emperor who had convinced himself he was wearing new clothes but was actually naked. Professor F is adamant that he is debunking or exposing theist-scientists yet he uses other people's data and papers and has nothing of his own. Scientists like Dr. Tomkins and Dr. Tour always provide mostly their personal findings during discussions, while Farina solely rely on other people's arguments. The current problem with this reliance on science is the disturbing increase in science fraud (discussed in another section) where one professor sated that **academia has been broken**.

Some atheists had asked: "why do bad things happen to good people?" Well, in reality, it's because of a myriad of factors that contributes to any event (see the last section of this book) but I can also offer another personal perspective, and that is: the devil is **one (just one) of the main causes** of many bad things that happen on Earth. People have different views on any matter though but if one read, discuss and do research on the following topics, they would find that one of my approaches can offer some explanations to this question:

70

71

1. Evil exists.
2. God's craftmanship gave humans free-will to decide.
3. God sometimes intervene – when and where one might ask. One can ask HIM when you get to heaven. So be a good person so that you can get to heaven and get the answer yourself, because from my personal perspective, I don't agree with any of the answers and explanations I have come across during my years of research for my books.

It's interesting to note that atheists, during televised debates, often never blame the devil for any bad thing happening on Earth. Atheists, at least those that "lack belief" (different from those who don't believe in the Christian God) often argue that Christians need scientific evidence, scientifically researched peer-reviewed sources to substantiate their points of argument. Let me provide my perspectives on these two issues.

Evidence has resulted in hundreds of innocent people being send to prisons all over the world. The legal systems, most of which employ very intelligent people, relied on various evidences to prosecute and convict these people. Scientific evidence has contributed significantly to these cases in the past several years. When forensic science with its DNA and genealogy approach entered the legal system a few years ago, the communities were dumfounded when they found out that hundreds of seemingly nice people, referred by some writers as 'fine next-door neighbors", were actually heinous murderers. At the same time, hundreds of innocent relatives and friends have also been wrongly convicted all because of "scientific evidences". According to Samuel Little, one of Americas most notorious serial killer, he confessed after many years being locked up in prison because he found out there were many innocent people locked up in prisons (using evidences) because of the killings **he** personally committed. Law enforcement and Samuel estimated that he killed about 50 people. Thanks to Mr. evidence, many innocent people have been jailed for Samuel's crimes. Recently, the US National Institute of Standards and Technology, US Department of Commerce, announced that the so-called scientific evidence of "bite marks" is invalid. This is another pseudo-science debunked. Sadly, many prisoners are in jail, some for more than 30 years, because of this scientific evidence. Congratulations to Richard Dawkins and his "science-based evidence" beliefs. Fortunately, the Innocence Project has exonerated about 30 prisoners so far. Evidence had resulted in thousands of broken families and hundreds of suicides due to wrong convictions in the US in the past two decades. A total of 49,449 Americans died by suicide in 2022 according to the US Center for Disease Control, and some of these incidences can be blamed on the false evidences used in trials.

Atheists' reliance solely on scientific evidence is sometimes foolish and the weight of this reliance is overwhelming. In recent years there has been an increase in science fraud according to some scientists and professors. I don't expect the atheists to believe my perspective on this but they can do some research on their own on the impacts on this reliance, via internet and also personally talk with those innocent people impacted by evidence. Sometimes evidences and interpretations lie, but facts don't lie. There's a difference. People are sometimes framed and false eye witnesses result in years in jail for some innocent victims. In early 2023, it was reported in the US that a male, Lamar Johnson who served nearly 30 years of a life sentence was released as he was innocent and wrongly convicted because of the use of "evidence". Some members of his family have passed away persuaded that evidence rule. Absolute verifiable evidence is needed in court, not just "evidence". Many "monsters" who have been free for 10-40 years have finally been identified through genetic genealogy in the recent past, but have been living pleasant lives outside of jail. A man interviewed for this book believed that prosecutors and law enforcement people involved in these types of false cases (and scammers) are good candidates for execution. I disagree though. Before being caught, they were deemed innocent because there were no **evidences** or because of weird evidences. Just because humans can't find evidences of God (there are however many evidences available if one does research on current discoveries, read powerful testimonies and several related books, and also watch many podcasts and related videos) doesn't' mean there's no Christian God. Discussing this issue, Pepe Seiuli, a local man offered this funny comment. Atheists may try forcing oneself not to breath for about five minutes, after which he will definitely believe in God.

As for peer-review scientific papers, these are **generally trustworthy sources** often utilized by professionals. However, the public should also be aware of increasing flaws and fraud in these scientific papers. In 2023, the President of Stanford University stepped down after acknowledging that he should have been more diligent in seeking corrections regarding his reviewing of some peer-reviewed papers. It was found by a Board that the research on one of the papers contained "various errors and shortcomings", but the Board didn't find evidence that the President was aware of the lack of vigor in reviewing said papers. This is not an isolated case and I know additional research have uncovered many more. For example, Diederik Stapel's audacious academic fraud and Hwang Woo-Suk fake stem cell research. Dr. Diedrik Stapel, a well-known scientist of Tilburg University, had more that 50 of his papers retracted because of science fraud. Scientist Jan Hendrick Schon had his Ph.D. revoked and Haruko Obokata of Japan made fraudulent claims for her "stimulus-triggered acquisition of pluripotency (STAP)". According to some researchers, "to make matters worse, 90% of allegations of biomedical research fraud are dismissed by responsible institutions without any faculty assessment or auditable record".[72]. Sole reliance on science in arguments against creation has become a farce recently and some universities are reluctant to investigate these allegations because it would **tarnish their reputation, affect professors' tenure and consequently lose some of their funding.** This is similar to a

---

[72] *The essential need for research misconduct allegations audits.* Lisa Loikith, Robert Bauchwitz.

scam investigation in Cananda where some victims-institutions of a scam that spread to US refuse to assist investigators. This was related to a man name Sebastian and it was discussed on the program called Fifth Estate regarding the Netwalker ransomware.

Several co-authors of many peer-reviewed papers now seems to distance themselves when fraud is discovered. Lately, some professors have left their jobs because they were disgusted with the growing number of science fraud. To be fair to science, psychology also has its share of fraud, described by some analysts as "questionable research practices" like p-hacking. The SocialNeuro article "Why science fraud goes deeper that the Stanford scandal" plainly summarized this issue by stating that "In summary there is a prevalence of science fraud". In fact, Science Fraud goes deeper than what has been uncovered and this is one gateway where voodoo science can emerge and argue against Genesis 1. In the field of Behavioral Science a related controversy was reported by a Board from Harvard University relating to Dat Fraud but the alleged Professor has since suit the university and the drama continues. World expert in this field Dr. Elizabeth Bik has revealed several fraud cases within the science community but has also noted that these are just "a few" bad apples within the mostly fine collection of good scientists. However, many such fraud are still out there and many more will be discovered. For example, the founder of the Institute for Behavioral Economics Research (TIBER) was found to have faked over 50 studies[73], and this is not "a few". This phenomenon is, however is not new occurrence. About fifty years ago, some scientists were actually being **paid** to "downplay" the significance of some of their data (Sugar Industry and Coronary Heart Disease Research, Cristin E Kearns, Laura R. Schmidt, Stanton A -JAMA Internal Medicine -Special Communication) to fit the sponsor's narrative. This is according to Dr. Robert Lustig.

In relation to Alzheimer, the prolonged and costly journey to

---

[73] Pete Judo. "The scientist who faked over 50 studies"

get drugs to treat the amyloid plaques turned out to be based on fudged data. In the field of history, there is a similar phenomenon that I describe as "unfair and biased representations". One can watch a related presentation by Dr. Sam Hanes, a "de-constructionist historian". Some of these presentations have the ulterior motive of trying to sell books but contain several errors pointed out by analysts.

A misnomer that appeared in recent years is the labelling of the Christian God as the "God of Gaps, where atheists believe that Christians bring in God whenever science fails to explain something. This is a ridiculous concept and a misrepresentation of the truth. Humans are both physical and spiritual and therefore the whole and complete human persona or condition should be explained in both these realms. Unfortunately, several atheists are not bright enough to explain the spiritual part of the human condition, so they themselves should be labelled as "**people of the gaps**". Christians on the other hand can use both science (read John Lennox and Stephen Meyers articles and books) and Christianity (I'm not sure of the other religions) to explain both the physical and the spiritual human domains; the **complete** human spheres of existence. Therefore, there is **NO** gap in Christian faith, it's a **complete** explanation. The word gap therefore belongs to the atheists not the Christians. Because the phrase "God of the Gaps" has been erroneously used by atheists numerous times, I will now use "people of the gaps" in the following sections to describe the atheist in case they forget who they truly are.

Professor Lawrence Krauss noted that many scientists like Einstein would mostly sit around their offices and think up various theories and extrapolate on available data, but that he would prefer them doing **actual experiments** like he did. He further explained that, as a result, the public gets what he called "intellectual masturbation". This is one of the reasons why science gets the "gaps". Science should advocate the use of experiments, especially those they personally perform, and not rely too much on textbooks (like Professor F.) because this will slow down the advancement of knowledge, dims reality and materialize gaps. World-renowned physicist Dr. Neil Turok stated that he describes theorists as "imaginary people" and experimentalists as "real people". Lets be real my friends and get off television debates if you don't have your own supporting evidences.

Atheists who believe in the scientific method should start their investigations by stating a hypothesis like "the Christian God doesn't exist". Their next step is to get the required data to prove their hypothesis. If they can't get the data they should backtrack to their hypothesis and maybe infer that their hypothesis **can't be proven**. This is ok since science is progressing?.

For those that see themselves as "lacking in belief"[74], (some groups of atheists have this "belief", like my friend Dan Barker) their world is a void space that is aimless and lacks fulfillment since it has no goals that may bring joy when reached. They normally hide this demeanor according to some who have left the group. Hannity, a newscaster of Fox News refers to well-known atheist Christopher Hitchens as a "jackass, hateful and mean human". I'm not sure about this statement though, but I think Hannity was referring to Hitchen's comments regarding the passing of Reverend Falwell. Christopher also said that Islam is a stupid religion. (Talladega Tom. 2013 "Hitchens explains Islam and why it is so stupid"). I'm not sure about what he meant, but he seems to lack respect. Maybe the Muslims should give him a visit so he can explain his comments.

Arrogant atheists who have made careers out of public televised debates against Christians may reconsider being in the limelight since some new scientific discoveries might appear in the future and contradict their views and basis for their arguments. Additionally, prominent scientists themselves contradict each other. For example, Dr. Jason Lisle (a Christian) contradicts atheist Neil DeGrasse Tyson on public television – both are astrophysicists. This group relies on science and yet they disagree with each other. Lost souls. They also argue that they depend on reason. But the Christian God is

---

74

God and doesn't depend on reason. Now they have absolutely no grounds for any of their arguments. Lost humans with no souls, maybe. Neil deGrasse Tyson, in one of his presentations said that he doesn't believe in Intelligent Design by God, then he provided a photo of a human female with a "patch" covering her genitals. He then mocked the location/appearance of the female genitals. Maybe I misinterpreted his statement but his is the only belief by Tyson that I truly agree with (or NOT) in that "female relatives' genitals should be located on their foreheads, not between their legs as created in Genesis". Some said that it takes intellectuals to come up with the dumbest ideas.

One final thought regarding this topic is the question asked by many atheists, that if God is good, then why are bad things still happening on Earth. I did provide an explanation to this question in another part of this book, however, when I asked one of my village chiefs about this issue, he told me that there is an experiment that can be done to prove that God's Earth was initially "good", as mentioned in Genesis 1:31, but it was man that started ruining it. He looked at me and said, just transport all the atheists, and the animals (murderers, scammers, conspirators, pedophiles and serial killers) to another planet. (The label 'animals" is found in scores of podcasts, documentaries, testimonies and interviews all over the internet, so this is appropriate). After twenty years, people left behind on Earth will experience a very pleasant atmosphere in their communities and with much less crimes. The world will then return to its Garden of Eden condition. I replied that this is not practical but he had already walked away to work in his *taro* plantation, while saying "it is the only answer to their question", and that he's considering letting all the televangelists who preach the Prosperity Gospel join the expelled group.

Let me end this section on a quote from the King James version, Psalm 71: "In thee O Lord do I put my trust, let me never be put to confusion"

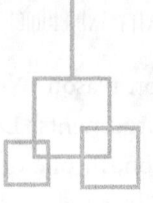

Prioritizing Lives in These Islands

After much research, interviews and drawing from experience, I was able to prioritize the following circumstances or considerations in relation to the practical lives and realities of our local Christian communities. This list is my own personal summary and conclusions of findings but will definitely conflict with some of the views by other indigenous Samoans including a couple of pastors I interviewed. The following is a list of four of these that I feel are the general feelings of our local people in relation to what their lives' priorities should be instead of what is required by the church and the culture. I must confess that this analysis is biased because I recorded the data after I discussed various related issues with the respondents and this may have influenced their perspectives at that time.

Priority 1:

Personal health: Parishioners, especially those of the Pentecostal churches won't be able to raise their hands and clap during their normal praise and worship services (*viiga*) if they have *loomatua* under their arms. Untitled men who have neglected their health conditions and have their limbs removed due to diabetes won't be able to assist in construction, repairs and maintenance work needed for church projects. Parishioners with various medical conditions will generally be unable to assist and participate in church events. So, this has to be addressed **first** before one considers spiritual needs. One can't love God with ALL their heart and mind if one has constant physical, and mental pain. Take care of this priority first then your mind is free to praise God.

Priority 2:

Children's education: According to government surveys, the gap between well-off families and the poor is mainly the result of education. Families that push their children to have a good education eventually have financially-stable homes. When a church has members that are financially stable, the church will grow and members of the church as well as the village will benefit.

Priority 3:

Church: After satisfying or taking care of the above items, and if there are resources (time, money, expertise) still available, then and only then, parishioners should consider providing these to the church. God doesn't need your money; He owns all the money in the world but He just want your soul and spirit.

Priority 4

Culture and the extended family:
This should be the least important item and yet several families I interviewed confessed that they placed this as their number 1 priority. Some of the families I interviewed confessed that because of the shame they will experience if they don't contribute to *fa'alavelave*, they have to provide money and *ietoga* for cultural events even if their children have to walk barefoot to school. I have some experience in

this item since a few times, when I walked to school (Samoa College 1969-1973) without any footwear and had no breakfast 80% of the time!.

Let me close this section with this warm feeling. It's personally fulfilling and gratifying to view and listen to so many Science Professors and other professionals trained in the sciences and religion arguing for the Christian God through various media platforms today. Because of this observation, I think I will now raise my bar and refuse to debate any more atheist who is not a Science Professor. Furthermore, anybody - atheists, theists, scientist, pastors, members of the public, who have questions regarding science and religion: I don't have all the answers obviously, but I've presented sufficient sources in my Footnotes and elsewhere to point anyone in the right direction. My contact information is in this book (About the Author), or just examine the several presentations by Professor John C. Knox, Dr. Tours, Spike Psarris[75], Dr. Jason Lisle[1] and others available on the internet. These fine scientists have provided arguments against atheists and at the same time satisfying 1Peter 3:15: KJV. "But satisfy the Lord God in your heart, and always be **ready to give a defense to everyone who asks** you a reason for the hope that is in you".

---

[75] Physics disproves aesthetic cosmology: New Creation Network. Apologetics Symposium. 2016.

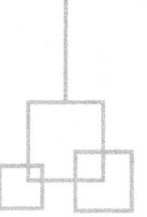

The Paranormal

I n American Samoa, a parishioner of the *Faleniu* AOG church visited a female friend at her home when the friend failed to appear for several weeks. After knocking on her door, her Filipino friend, a Catholic, opened the door and the visitor was taken back at was she saw. She later described her friend's deranged face as resembling the faces of demons and the devil she used to see in horror movies. The room was dark but had several statues of Mary with candles burning all around the room. The visitor asked her friend what was going on and she replied that she was praying to Mary to get rid of the demon spirits in her house. The visitor immediately offered her help by asking her friend to pray with her not to Mary, but to Jesus Christ. Immediately after their prayer, the friend's face returned to normal and she joined the visitor in a prayer acknowledging Jesus Christ as her savior. Readers are very welcome to contact this author so they can talk directly to all the ladies in this story including Rev. Lalagofa'atasi To'omalatai Sanonu, the pastor of this Faleniu AOG church. I like it when I could get the doubters get stories right from "the horses mouth".

The Catholic Church has always been the church that dealt with the phenomenon of demon possession and has priests specializing in exorcism. The phenomenon has declined in the last decade in these islands but has increased overseas. The global rise in demonic possessions prompted the Catholic Church, during this period, to reopen exorcism schools to train priests in the Rite of Exorcism. Surprisingly, a young Sister Ann (although forbidden to perform exorcism) and Father Dante were trained recently by Professor Colin Salmon in this field. This is another example of how women are increasingly recognized in churches. Some Pentecostal churches are now also helping with exorcisms.

A related phenomenon that I found very interesting are the messages related via psychics, from the spiritual world. My research and also experiencing this phenomenon had made me a believer in many of these paranormal events. Readers may research this topic on the internet where many well-documented examples of psychics had assisted law enforcement in solving many homicide cases in the US and Britain. For example, a psychic in New York was deputized by the Police Department when he correctly located a missing boy after more than 300 people unsuccessfully searched for the missing boy in 1975.

Explanations of how mediums and psychics sometimes identify homicide suspects and consequently solve many homicide cases should also be explained by atheist-scientists from their science perspective. Unfortunately for them, these phenomena can't be explained by science and yet used formally by law enforcement in several countries. Law enforcement officials have formally solicited the assistance of mediums/psychics, for several years in Britain and other countries, when scientific processes fail. There's definitely other unknown phenomena beyond science!. Concepts like the human soul, spirit and consciousness. Chief Lolo from my village can explain these concepts if one visits the district of *Falealili*. But for now, the atheists can continue to stumble by debating theists that often appear on podcasts and varies presentations on the internet. Evolution and Creationism are mostly irreconcilable views according to some scholars. Public debates on these issues are mostly farces for public entertainment. But these discussions may also be beneficial mental exercises for "thinkers" to facilitate what I described as "dramatization of reality". Christian scientists have also explained how the Bible can explain some of the issues that have baffled scientists.

More than 30 years ago, a male psychic relative of mine named Pule used to assist the police in Apia solve a few cases regarding stolen money. He used to live across from the Saint Joseph College school. Every time Pule points the police to a suspect (he was 100% spot-on) and the case is solved, he would experience physical pain that would leave him bedbound for about a week. A couple of ladies on Tutuila also had this gift and were able to solve a couple of cases regarding stolen money[76] in the late 1980's. These mediums are always willing to help but they sometimes are reluctant because of the aftermath pain they will subsequently experience. A few psychics in Britain reported on national television experiencing similar physical pains while assisting police in various investigations. Again, I whole-heartedly believe in the paranormal. Sometimes the information from these people is incorrect but according to my research; it's the limitations of the mediums (human psychic) that cause this flaw. I think this phenomenon is also mentioned in the Bible.

I like these types of paranormal stories because these are like Samoan Legends where the facts mostly don't add up but there is also the possibility of some of these being true! An example is the Samoan legend, (that is now absolutely true to me) is the story of a female ghost (*ilamutu*) named *Telesā* that is supposed to hang out in a certain part of the village river, a few yards from the piece of land called *Sepolataemo and Mulinuu*, at the village of *Lufilufi*. I grew up in this village when my late father Rev. Aitaoto Seiuli Fuimaono was the pastor of the LMS church in that village. One fine night when the moon was shining brightly, I was outside our *fale* Samoa late at night when I saw a female with long shiny hair, back towards me, sitting on the very large piece of rock that was in front of the pastor's house[77]. A few seconds after I saw this, I couldn't see anything else,

---

76

77

I was totally blind. I called out to my brother Muti and he came out and led me back to the *fale*. In hindsight, I'm still puzzled why he didn't ask me what was going on. The event didn't make sense to me until several years later when I heard stories of *Telesā*, the female spirit that frequent a certain part of the *Lufilufi* river. She would often sit of the same rock in front of the pastor's house, according to the elders of the village, but most of them haven't actually seen this. Maybe their ancestors saw this and the story continued through legends. Since I've personally experience this, it shouldn't be a legend anymore. I've experienced this and this should now be a personal fact. (The Cook Islands has a very similar legend about a hauntingly beautiful moon spirit name *Taakuva*, who also has beautiful long hair). From this personal experience I have concluded and **without any doubt** that the legend IS TRUE and any reader can contact me and I'll pay for dinner while discussing this event or even take them to the village where this happened. This is crucial to me as I try to convince people that some legends maybe be true. David Montegomery of Harvard University in 2015 stated that "oral legend can be true". Most readers would prefer that sources for this kind of phenomenon should come from prominent members of a community to be of any value. Well, in 2023, Mexico's President Andres Manuel Lopez Obrador posted a photo on his social media accounts showing what he said appeared to be a mythological woodland spirit similar to an elf, a spirit mentioned in Mayan folklore.[78]. For this book, the phenomenon of *Telesa* is similar to the story in the Book of Enoch, where females were abducted by beings from outer space. Of course the Book of Enoch is not in the Bible but then Enoch is mentioned in Genesis.

Like the phenomena of UFO with its thousands of photos, interviews and other sources, the study of reincarnation is also a solid field of study. The University of Virginia has more than 2,000 documented cases of reincarnation cases where several people (mostly children) have demonstrated that they actually lived in the past, including the fascinating case of James Leiniunger. Many other documented cases with personal interviews include the cases of Ryan H, Nicola Bird, Carl Eden Carr, Cameron Robinson, Pam Robinson and many others. After a few years of research for this book, this author believes not only in UFOs but also in reincarnation. Science rules, according to members of our *aumaga*, but always under the decree of the Christian God. Readers may also want to read "Little Green Men" by Dr. Hugh Ross.

A little humor on this topic was provided by one comedian when he said that he doesn't care much about muslims suicide bombers because they can do it once, but he's scared of Buddhists suicide bombers because they can do it multiple times through reincarnation. In these islands, White Sunday *to'na'i* (special Sunday feast-brunch for the children) is a very unfortunate day for local chickens and pigs. Reincarnation is therefore not favorable locally.

The phenomenon of near-death experiences has been well documented and according to some researchers, there is a lot of proof to this phenomenon where patients actually died and later described in detail what happened after they were clinically dead[79]. Even some atheists have documented experiences of their near-death experiences as described by D. Bruce Greyson in his book "After". Readers may also read the book "Author and Near-Death Experience" by Anita Boorjani. A researcher, Dr. Janita Holden who reviewed 93 reports of out of body experiences found out that 92% of those accounts were accurate. Doubters may also examine Dr. Mary Helen Hensley's testimony on "Anthony Chene Production" 2023. In addition, more supporting information may be found in texts by Sarah Hinze and Dr. Raymond Moody's "Life After Life". According to Professor John Fisher, 90% of documented cases are positive or pleasant experiences and **many can be corroborated**. Readers interested in this subject may also be interested in the related

---

[79] Dr. Dodge Rea. The change paradox. TruStory FM.

phenomena "Pre-birth experience" and "Pre-natal Connections" as described by a Sandra Demchuk. This phenomenon is real and also brings up the idea of consciousness being a function of not only the brain but of some external force/energy.

There are, however, problems with the researches on this topic. For example, some samples are biased; there were also cultural influences; ineffability, reluctance to reveal experiences and inconsistent protocols. These finding were from more than 40 years of research.[80].

The use of high-sensitive technology today by several investigators have also provided scientific proof of the existence of ghosts and spirits within various communities. Paranormal, demons (refer to Mark Chapter 5) and ghosts are real and confirmed in these islands. I (and many others), have personally experience the presence of ghosts and sprits of the dead at Manu'a more than thirty years ago when my late father was the pastor for the LMS church in Tau. This phenomenon is true and real. Readers can research this topic on the internet and may learn of the authenticity and well documented messages from the spirit world.[81]as high-tech equipment document their presence. So, when scientist like Michio Kaku and Neil deGrasse Tyson said they don't believe in human spirits after death, it was contradictory to their science convictions, because there are now well-documented scientific proof of ghosts.

A personal story that happened to me in June 2023 can add some flavor to my belief in this concept and readers can contact the lady (Agnes Sesepasara of Hawaii) involved. Agnes called me from Hawaii early one morning in 2023 and she was crying. The bedridden lady said if I could do her a big favor and I said yes. She said a few minutes earlier, her husband was standing beside her bed and when she tried to touch him, he disappeared. Her husband, Lepe Sesepasara, passed away more than ten years ago and he is buried

---

[80] University of Virginia. UVA Engagement. 2022

[81]

in Tafuna, American Samoa. On that morning, her sister, who was her sole caregiver, left to live with her own children, leaving her alone. She struggled to call her son, grandchildren, and relatives – which included me and my children. That was when she called me. She requested that my daughter Renee, go and place a flower on her husband's grave to thank him for visiting her when she was all alone. I did her the favor and will continue to believe in Samoan human spirits roaming these islands, until the day I die. Also refer to my other personal story about my Grandma Sina, who visited me after she died more than 40 years ago! Traditionally, Samoans truly believe in ghosts and spirits from the afterlife. Interested scholars can also talk with the chiefs of Tufutafoe village in Savaii, who are the authorities and experts on this subject.

During this period the ancient practice of "*fono ma āitu*" ("confer with the ghosts") was reawakened by Samoan Culture experts Tui Atua, Maulolo and other scholars. This event portrays the practice of providing a request to the spirits or traditional gods, to bless or allow a certain event to proceed. A coconut is left in a dark *fale* in the middle of the night and the *aumaga* (untitled men) would check this coconut at dawn. If this coconut is later found to have been drunk, then this will be taken as proof that the request is approved. This should be taken as scientific proof as this practice has been **repeated** several times by many chiefs and people in our community with **well documented video** – a requirement of the Scientific Method. Therefore, from the *palagi* perspective, this traditional belief should be true. During a relatively recent reenactment of this event, a pastor participated – a modern addition to this old tradition. In December 1986 during the bestowal of the *papa* and title *Tuiatua* at the village of *Lufilufi*, the event started with a church service, another example of a modern addition to this tradition, which had no religious element before. (I think this was the church I used to attend several years ago when my father was the pastor of the LMS church in that village). A personal experience in the addition of a Christian concept into traditions happened when I was bestowed the High

Chiefly title *Fuimaono*. The ceremony started with a church service and the blessings by church pastors of the different denominations. Remember that these islands were settled around 800 BC while Christianity was just introduced in 1830.

Moreover, Samoan orators (*tulafale*) often refer to a place in *Falealupo* called *Pulotu,* in their speeches[82] during funerals. *Pulotu* is the place/cave where Samoans believe all the spirits of local people go to when they die. I'm not sure I would like to go there. The legends from Fitiuta, Manua also tell of their own similar cave (pers. comm. AOG Rev. Scanlan) and from theses legends I formed a personal *alagāupu* (traditional saying): *O ou paia Samoa mai le ana se'ia pāia le ana"* - salutations beginning from one cave to the other, covering the whole Samoa group.

For the skeptics, there is no better proof of the existence of friendly spirits of the dead than evidence that originates from members of the clergy. The notable pool at the *Piula Theological College* is the location of several related stories from the students and teachers of this prominent Bible College. One very interesting story tells about female students and groups of *faletua* who normally clean up the famous pool and its facilities before tourists and visitors arrive. The ladies were supposed to clean up one day when they discovered that the place has been well cleaned. They thought it was one of them that did the job then they left. Later they found out that none of them actually did the cleaning. The occasional occurrences of similar "cleaning" incidents made the *faletuas* believe it was the friendly ghosts that did the work. Interested readers can read about these awakening stories in the book "God is Samoan – Dialogues between Culture and Theology in the Pacific" by Matt Tomlinson[83], Pacific Islands Monograph Series 29. 2020 UH Press.

During the great Lava phenomenon on the island of Savaii

82

83

(1905-1911), the Lava went around a church that was right on its path. In 2023, the great fire in the island of Maui, Hawaii, burnt down most structures in Lahaina but a church that was in the midst of the fire was unscathed. A powerful tornado that was called the Beast, damaged many parts of the 59 US States in 2021, just before Christmas. A family living in Western Kentucky decided to run to their church to wait out the twister, which was about a mile wide with winds up to 190 miles/hr. After the tornado, the family returned to their solid brick house and it was completely destroyed. The man was asked by people of the community how they made the right decision to leave their home and made a run to their church. The man replied "the good Lord told us to go to the church".

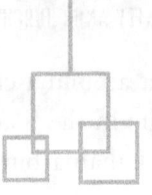

Gender Equality

A comment which appeared on the internet stated that "the U.S. Constitution was framed by some slave owners whom said all men are created equal". I'm not sure if there was sufficient research done on this statement but it's giddy in a way. With Samoa electing its first female Prime Minister during this period, it was a historical moment for all female leaders around the world as shown by the many congratulatory messages received from overseas countries after the national general election. The new Prime Minister, Fiame Naomi Mataafa also brought with her several women for top positions in her new government. The new government was brought into power through its political party called *Faatutua i le Atua Samoa Ua Tasi* (FAST) translated, "Have Faith in God, Samoa is one". This motto clearly reflects the expressive relationship of FAST to Christianity.

The FAST administration of Samoa started addressing the issue of Gender Equality in its initial years. The case of Ropeti Sione who escaped to Australia in 2017 and who changed his name, had child support obligations and a Stop Order against him, brought this issue to the forefront when Sione was deported from Australia. Advocates of Gender Equality argued that this should never had happened in the first place. The FAST government's new funding for NGOs to combat gender-based violence will assist this movement.

An international global volunteer movement called the

Soroptimist, which started 100 years ago in May 1921 in Oakland, was to donate 100 bicycles for girls and women in Samoa in 2022. The bicycles were donated by Australian residents. The mission of the project was to transform the lives and status of women and girls through education, empowerment and enabling opportunities. The movement advocates for human rights and gender equality.

Generally, there is no equality in life. Humans are born into this world with **unequal** resources and capabilities and therefore unequal situations and outcomes in their lives. Some people are born of wealthy parents, some born with various disabilities, some with high/low IQs yet some are born gay, and when humans start their life's journey at different starting points, and with unequal resources, inequality becomes a fact of life. On the topic of Affirmative Action, I believe I this approach when applied to education but not when considering candidates for a job.

Down the line, differing relationships and the environment (social and physical) impacts one's life's journey. This further increases the inequalities in people's lives. The concept in the Bible about everyone being equal is just a spiritual concept and not in the atmosphere of reality. Jesus chose just men to be his main disciples; Eve was created after Adam and was not his equal but just a companion. A few women, later in Biblical times however contributed well to spreading the Gospel of Jesus Christ.

In these islands during White Sunday, the children would often say: *Ia manuia le Auauna a Le Atua, faapea matou le fanau soifua"* (God bless the pastor and us children), at the end of their presentations. In the past two decades, the following addition to these final words was introduced. *Ia manuia le auauna a Le Atua, fa'apea le faletua (*God bless the pastor **and his wife**). *Faletua* refers to the pastor's wife, and the addition of *faletua* reflected the new idea of including females in societal salutations, a foreign concept to the traditional Samoan Christian worship since 1830, and a push towards Gender Equality through Christianity. With the election during this period of Samoa's first female Prime Minister, there is no turning back.

If I may insert some interesting observations regarding the law (and general inequality) in the US (American Samoa is a US Territory). The public have always believed that the law in the US treats everyone equally. Not. For example,

- If the FBI (the US Federal Bureau of Investigations) searches an attorney's office, the Attorney is given the opportunity to request a special assessment to evaluate the legality of the search and determine the relevance of documents seized. This privilege is not extended to ordinary people even though there might be documents that might incriminate the person searched.
- When a former President was indicted in 2023, he wasn't initially required to take a mugshot **like everyone else**. He didn't in the first convictions but later did had one.
- The US Supreme Court doesn't have a Code of Ethics (that was later changed in 2023 but unfortunately it lacked enforcement according to one commentator) and refused to examine alleged illegal dealings by one Supreme Court Judge. Oliver Diaz, a retired Mississippi Judge said that the US legal system is a "2-tier system," where the rich are able to hire expensive attorneys, investigators, witnesses and have other resources, while the poor don't afford these luxuries and consequently end up in prison (referring to Michelle Bryan's case). He said most of the prisoners during his time in court are poor people. There goes equality in the US legal system, out the window.
- Also consider the 2008 US financial meltdown where the top banks and Wall Street executives were not jailed (they just caught a few "small-fish") after they ruined the lives of millions of people. To some political analysts, it's a national legal disgrace.

- Maybe the US Legal system should revisit the stupid Qualified Immunity concept according to one Legal scholar (e.g., James King case) and the unfair Private Equity concept.
- There are luxury prisons in the US for those who can pay. "Pay to Stay" jails have certain perks. Maybe there should be such privileged prisons in these islands so those prisoners who complain about food and certain practices can have similar options.
- Pedophile Jepstein Epstein's wasn't properly prosecuted after **four consecutive administrations.** He later committed suicide and his partner in crime (Maxwell) was jailed. Names of several prominent people with ties to this man was revealed in around early 2024 and included a former US President and a British Prince.

These are just a few examples in the US but much more similar and horrible events occur in other countries, with a few in these islands.

During this period, MSNBC's Lawrance O'Donnell stated that one of the US Justices "is the most corrupt member of the U.S Supreme Court". A legal scholar said that Anita Hill was right all along, a relevant case in the past. Furthermore, according to MidasTouch 2023. Ben M. "Make no mistake, a few US judges are both corrupt and sloppy". I don't think this refers to Judge Aileen Cannon?. (Glen Kirschnek. 2023) and the judges are just fine.

During Mother's Day 2023, the Samoa Museum at Malifa exhibited archival pictures of all Mau leaders as well as Mau Women's Committee. During the Mau movement, the Mau Women voices and activities were percussive and strong. A book "The Women's Mau", included a letter dated August 23, 1930 from Rosabel Nelson to Olaf Nelson which stated that the Mau Women had a membership of over 8,000 and had received assistance from the Women's International League for Peace and Freedom. I wish my late mother was here to read this section.

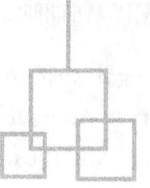

Religious Education

Discussions concerning the establishment of the Malua Theological College (MTC) began in 1840 and a formal agreement was reached in February 1844. The Malua Theological College was later founded in September 1844 by Rev Charles Hardie and Rev. George Turner of the London Missionary Society (LMS). The vision of MTC was "For Jesus and His Church". The area that MTC was build on was formerly known as *Maluapapa* and its now commonly called Malua. The College offered its first courses on September 25, 1844 with only 25 single-male students. Married students and their wives were admitted in 1846. After twenty-five years, 1143 students had graduated from MTC according to Rev. George Turner.

My father graduated from MTC top of his class and later served as an LMS missionary in Papua with my mother. His name is included in the names of late pastors etched in the monument at MTU. A sister and brother were later born to them in Papua Niu Guinea, both have passed away. He also served as pastor for the Vailima, Tau Manua and Lufiflufi LMS churches and was the Church Treasurer serving out of the *Maluafou* compound, for a few years.

In 1850, the London Missionary Society (LMS) established a boy's school at *Fagalele Leone,* American Samoa. The goal was to successively enroll the schools' graduates into Malua Theological College (MTC) in Upolu, Samoa. The LMS girls' school at *Atauloma*

(American Samoa), established in 1900, and Papauta (in Samoa), founded ten years earlier, were established to prepare girls to become LMS pastors' wives (*faletua*) upon graduation. My mother was a student at *Atauloma* and she eventually married my father after graduation.

Most Samoans have since subscribed to the Bible texts that is believed to have been first written, in part, around the 10th. Century B.C. with the most recent chapters scribed in the first two or three centuries A.D. In the US, the Supreme Court in 2022 ruled that state programs providing money for public school tuition cannot exclude schools that offer religious instruction.

The five CCCS Colleges celebrated a milestone in 2021 with the launch of a new project to "enhance the knowledge of the student in terms of technologies and the use of Zoom". The project acknowledged the funding from the Council of World Mission (CWM). The project was launched at Maluafou College while students and teachers from the other four CCCS colleges joined in the celebration via Zoom.

In late 2021, CCCS anticipated that the Malua Theological College, will offer the Masters in Theology course in 2022. During the annual graduation, College Principal Reverend Dr. Vaitusi Nofoaiga said the course was to have been available in 2021 after it was passed by the Church General Fono (*Fonotele*) in May 2016. However, there were details that needed to be completed such as a library to serve the students. The library was funded by the former Malua students, *Malua Tuai* at a cost of $50,000 tala. The graduates at this time included 28 students with one earning a Diploma in Theology, 17 with Bachelors in Theology and 10 with Bachelors of Divinity with Honors. Most of the students are Samoans from Samoa and overseas denominations and only one student from Tokelau who graduated with a Bachelor's degree in Theology. Two from the Kiribati Uniting Church graduated with Bachelor of Divinity degrees with Honors. The graduation was attended by the Head of State, the Prime Minister, and Ministers of Cabinet.

Other than a few who are continuing their theological studies under church scholarships, most of the graduates will now wait to serve when they are called by the congregations that need a new pastor. The Congregational Christian Church of Samoa remains the biggest denomination in Samoa, during this period.

The Malua Theological College and Piula Theological College completed the initial steps towards the international recognition of Samoa Qualifications in 2022.

Rhema Bible College in Samoa graduated about 83 students during its 2019 - 2020 ceremony. More than 100 students graduated during 2021 and was the largest graduating class in the college's history. According the Director of Student Affairs, 2021 had been a year of significant expansion and was the 24th. School year where 114 students graduated. Ministerial Diplomas awarded were of Level Five accreditation according to Samoa Qualifications Authority (SQA). Around September 2023, their vessel *Uttermost Witness* visited American Samoa and I had the privileged of providing coconuts and breadfruits for the crew for a few weekends because the vessel was moored about 100 yards from my home in Malaloa.

Towards the end of 2023, the RHEMA Bible Training Center Pago Pago completed its first year of ministerial training since opening their classroom doors on August 1st. A total of 11 graduates from different backgrounds and some from abroad received Certificates in Ministerial Training. The students can return to complete this 2-year program in January 30. This is a 2-year inter-denominational ministerial Training Center that feeds into 1 3-4 year program at their Apia Center.

During this period, the first ever Professorship in Biblical Hermeneutics at Piula Theological College was bestowed upon its College Chair, Reverend Dr. Mosese Mailo at the Mulivai-o-Aele, Faleula temple. The Reverend is only the second to be conferred the title of Professor in the 150-year history of the Methodist Theological College. According to Mailo "A professor is not to be worn as a crown, but to serve; to run and work as untitled men, to serve the

Church through faith in God,",. The decision for the title Professor to be conferred onto the principal was decided on July 15, 2021, based on the recommendations from the *Piula* Committee. He graduated with a Master in Theology with Distinction from PTC in 2003, followed by a Doctor in Philosophy from Birmingham University in the United Kingdom in 2008. In 2019, he earned a Diploma in Management of Higher Education Institution from the Galilee International Management Institute in Israel. He has taught at the Piula Theological College for 15 years, including several years as the College's Principal. Rev. Dr. Professor Mailo is a former Chairman and a member of the South Pacific Association of Theological Schools (SPATS) executive committee, a former President of the Oceania biblical Studies Association, an Oxford Institution of Wesleyan Studies member and a Society of Biblical Literature U.S.A. member. He has been a Thesis Reader and Examiner in the Master Level and PhD level since 2014. Rev. Dr. Professor Mailo has been a Reverend (*Faifeau Fa'amaoni)* for 22 years when conferred this new title.

Early 2022 saw 24 students being accepted to the Malua Theological College after they passed the *Tautinoga* – a stringent interview process. The announcement was made at the CCCS at *Sogi* and was attended by many friends and relatives of the students. The school year was to start on February 6.

In 2022, the CCCS's Malua Theological College (MTC) celebrated a milestone with the unveiling of the publication *Samoa Journal of Theology*. Many church leaders and elders attended including members of the Elders Committee. The journal was compiled by the MTC staff to compliment the Masters of Theology program which started during the same year. According MTC, this was a first of its kind in the Pacific. MTC had been publishing a Malua Journal (in Samoan) annually but this new special publication is an international text that includes various compositions and multi-discipline academic writings from the staff. (I'll offer a copy of this book to MTC in memory of my father, a former MTC student).

Electronic copies will be available on MTC's website. The journal has been recognized internationally and will be published annually.

The Moamoa Theological College had a major fundraising in 2022 to upgrade its facilities. The 145-year-old college needed about $3 million for the upgrade. The institution was initially located at Savalalo, moved to Lalovaea and finally to Moamoa. The school's courses are accredited by the Samoa Qualifications Authority.

Rev A. Faaniniva, a resident of Brisbane, Australia, was one of the graduates from Piula Theological College in 2022. He graduated with a Masters of Theology.

In September 2022, nineteen students were ordained from the Malua Theological College after four years of Bible studies. Some of the graduates were from churches overseas. Various meetings were held prior to the graduation ceremonies including the Women's Fellowship meetings.

The Pacfic Theological College (PTC) announced in 2022 that Samuelu, a PhD student has successfully defended his thesis to the Board of Examiners. The PTC Board of Examiners consists of three professors from South Africa, Australia and New Zealand. Rev. Samuelu's thesis was: Decolonizing Grace: A *Faapalepale* Restorative Theology from a Samoan Perspective". In his thesis, Rev. Samuelu proposes an alternative translation of *faapalepale* which would decentralize the current view of grace.

The AOG Bible College in Tafuna, American Samoa had combined with the Antioch Bible College providing classes at the AOG main compound in Tafuna during this period[84]. This was an interesting event as this is the first time an AOG Bible College had combined with another Bible College that doesn't have AOG in its official name. First hand interviews suggested that this merge would never have materialized under the old guard of Rev Haleck and Mageo, but changes and adaptations may now be appropriate in these modern times…..maybe.

---

[84]

Twenty-nine candidates passed the Malua Theological College (MTU)exam in 2022 and 28 were ready to enter MTU with one planning to attend in 2024. The Entrance Exam was held in August 2022. This is a 4-year college and is owned by the EFKS, which has the largest number of parishioners in these islands, according to the latest government statistics, not including those in foreign countries.

There are religious countries overseas that limit education for their children to US grade 8 and lately (in 2022) the Taliban prohibited women from attending universities. Research into religion including Christianity is subsequently hindered by these actions. Education, however is not the only factor or goal for someone to be successful in life. People of the Samoan Islands should not let **schooling stand in the way of their education**. For those locals that are interested in studying Christianity, the resources now available on the internet is an excellent start and its free. Sometimes, the information available on the web is more in-depth, comprehensive and current than those provided by the local Bible schools. Again, the Holy Spirit is available 24/7 if one needs help. Wisdom on the other hand can be obtained from odd sources. During this decade, a professor speaking at a university commencement exercise professed that all his conceptions and perspectives regarding life was from his father who was a third-grade dropout and worked hard as a cook to provide for his family. Advice from those with extensive experiences in all walks of life can also provide immense rewarding, valuable and tested guidance to follow. A retired Navy Seal once said in his Commencement speech that if one wants to change the world, one needs to start by making one's own bed every morning.

Some events mentioned in the Bible seem to contradict Christian teachings and have been used by atheists to argue against atheists-scientists during televised debates. I personally believe that these unfortunate events were part of HIS divine plans. The prolonged trip from Egypt by the Israelites was filled with anguish, difficulties and constant struggles and it seems that God didn't really care about His people. Apparently, the prolonged exodus seemed to be God's plan to teach the Israelites about faith, belief and obedience.

In 2023, the RHEMA Bible Training Center Pago Pago (RHEMA BTPP) was established in American Samoa. The school will occupy space at the GHC Reid Building, near the Sadie Thompson Hotel where sinners like myself frequent to play pool and socialize. The school is part of the Uttermost Ministries Inc. and takes its name from the Greek work "Rhema" which means "a spoken word". According to the school, Rhema is one of two different Greek words found in the New Testament used to identify words from God with the other one being "Logos". The first Rhema school in the Pacific was the RHEMA Bible Training College South Pacific in Apia, Samoa, established in 1997 by Kelly and Pattie Duinick. RHEMA schools were later established in Fiji and Vanuatu. The organization has a 90-foot motor yacht call M/Y Uttermost Witness that is used for mission trips and as an educational tool. A total of fourteen local residents enrolled in August 1, 2023 with the majority being men and women in the 50-60 age group.

A Samoan Pastor, Professor Doctor Upolu Luma Vaai became the second Samoan to head the Pacific Theological College (PTC), Suva, Fiji during this period. Another Samoan Pastor, Reverend Faitala Talapusi also served in this capacity during 1992-1994. I personally know Faitala when we took courses at the University of the South Pacific, Fiji.

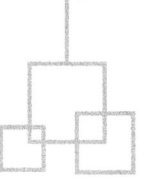

Suicide

This is one of the areas that the church can contribute to in conjunction with government agencies and NGOs. Statistics provided by the Samoa Police during this period, indicated that thirty-seven people committed suicide in 2021, and seventeen tried unsuccessfully. From January to October 2022, seventeen people committed suicide, and that the numbers have risen to 28 for 2022

This is a medical issue and therefore the medical professionals should take the lead against this problem while the churches provide other related support.

In late 2021, the Potter's House Christian Church in Vaivase addressed the issue of self-harm in a new drama production title "Suicide is never the Answer". The drama is a story of personal experiences and encounters with the issue of suicide. There were nine families in this church at this time.

In Alabama after the 2019 suicide of a local teenager, the community placed roadside signs that read "You are Worthy of Love", "Don't Give Up", "You Matter". Samoans should try to get basic information from the family members and teach the community on ways to identify the crucial signs to identify possible problems. For this, church support in needed in additional the government and NGOs expertise.

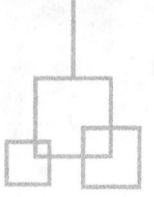

Preachers Controversial

This topic is included in my book because this author believes that preachers, mainly the **millionaire televangelists** in the US and elsewhere, should NOT be millionaires, accumulating so much money while millions of people are starving all around the world and even in their neighborhoods. For example, about 500 died from hunger in Sudan, including two dozen babies in a government-run orphanage in the capital of Khartoum, since the fighting erupts in the capital of Khartoum, around April 2023. This is not a new phenomenon. Preacher Robert Tilton had preached the disgusting Prosperity Gospel in the 1980's[85] asking people for certain amounts of money immediately. This would make a Samoan family struggling with only US$20/week to abandon church altogether. People trying to find the Anti-Christ may try looking at some of the pulpits of megachurches in the US.

I had personally observed a few preachers who had left two or three denominations and who had advocated for his previous denomination's doctrines. This had made me loose concentration when I attend church and listening to pastors preaching because they might be preaching a completely different interpretation the next time he preaches, in a different church. Readers may opt to do their own research on the scriptures, ask the instructors at Piula, Kanana

---

[85] Pablitos Way "The scam that made him $80 million per year"

Fou and Malua Bible Colleges in addition to requesting help from the Holy Spirit.

The fraud preachers preaching the prosperity gospel should sell their excess belongings – like their second and third private jets and expensive cars - and use the proceeds to feed the poor (refer to Luke 18:22). If they refuse to do this, then they are NOT preaching the Gospel of Jesus Christ. I'm positive that when any of these preachers read this, they will immediately twist and mis-interpret all Bible passages related to their false Prosperity Religion, which is what they have been doing for many years. An example is the verse: "Give and it shall be given unto you". These preachers should know (I know they already knew) that the verse actually mean "giving to the poor" if it's interpreted **in context**. It doesn't mean to give thousands of dollars to the church. They often twist passages like Genesis 8:22 and Galatians 6:7. I hope these preachers read this and fly over to our islands so the chiefs in my village can scold them. I know some of them, for example Creflo Dollar and Benny Hinn, who found out later that they have been teaching the false Prosperity Theology for decades, have confessed on national television that they were wrong (see documentaries and podcasts on the internet). One might deduce that these preachers have repented and finally "saw the light": but it was just the opposite. According to Biblical scholar Justin Peters, Benny Hinn went right back to his old ways. I've talked to three pastors with Ph.D. in theology and they agreed with me. These televangelists are disobeying the commandment of their Lord according to Mathew 19:21 and Mark 10:21, so they are not qualified to spread the Gospel of Jesus Christ. Mathew 25:26 clearly states what will happen to these types of people. Also refer to: KJV Galatians 6:7: Be not deceived; God is not mocked: for whatsoever a man soweth, that shall he also reap". Several local people I've interviewed, like watching these televangelists on television. Now they need to be well informed. If readers think that I'm too hard on these poor televangelists, watch Pastor Gino Jennings shatter these false prophets on national television, it's **much more brutal**.

Additionally, the Justin Peters Ministries has labelled a member of this group, Kenneth Copeland, the most "prolific false prophet ever" and that the Holy Spirit had never dwelled on him. The bottomline is that these televangelists don't pray for the people, instead they prey on the people – thievery. Ponder this, Jesus and the apostles **were not wealthy,** and Christianity is free religion and so no one should pay for it. These preachers should also learn how to hold things loosely so that God doesn't have to pry them out of their hands.

On a parallel secular perspective, in the US, this phenomenon is similar to the greed of Wall Street (around 2008) where ridiculous sub primes and derivatives-based processes negatively impacted thousands of low-income people while Banks and their CEOs pocketed millions of dollars. In Samoa, we call this phenomenon: *taufa'asēsē*. No Wall Street executive went to jail, and the US government charged only the "small fish". I'm not surprised that the Middle East countries refused to permit foreign government ideals into their countries, for these types of events inevitably come with democracy – which sometimes bring in negative impacts. According to students from these nations and one orator of Malie village, who is a Australian University graduate, they don't want to have the "untouchables" like the Wall Street executives, come into our islands and steal thousands of dollars from the public who are mostly low-income impoverished indigents. Furthermore, the practice of capitalism which often accompany democracy has some bitter dark sides to it. One is called inequality. For example, between 1977 and 2007, 60% of the U.S. income went to the richest 1% of Americans. (CAPTIAL. In the Twenty-First Century. Thomas Piketty).

In Brazil, "half of that country's money goes to only 5% of the population" and its biggest problem is corruption[86]. Bread and butter go together so as politics and corruption. In South Korea, a few families referred to as the Chaebols (The Untouchables) own

---

[86] Mood Side "The Dark side of Brazil" 2023.

and control most of that country's economy and politics[87]. This is not fair and definitely not Christianity. Secondly, contemporary economic analysis by some experienced company leaders suggested a new perspective on world economics, pointing out a few deficiencies in the capitalist system. For example, the long-standing assumptions that if wages increase, jobs must decrease and the price of an item is always equal to its value, are wrong. A presentation[88] regarding these ideas concluded that these faults has resulted in the continuous increase in the gap between the rich and the poor.

Furthermore, millions of people in the US don't believe in the democratic process of government. When a former President tried to illegally dismantle the Presidential elections results, millions of people agreed and supported him, including members of Congress, reflecting their opposition to democracy and the rule of law. Why is the US trying to introduce democracy into other countries when they, themselves don't believe in the democratic processes asked one commentator. A similar phenomenon also happened in Russia during this period where many soldiers were suspected of self-infliction to avoid combat. No wonder Philosopher Aristotle argued against some of the democratic concepts. Even a US Presidential Election denier was elected as the 56th. Speaker of the US House in October 2023. GOP Presidential candidate Chris Christie said that a former President made a controversial call last year for the "termination of parts of the US Constitution". The US democratic system has also permitted one offensive member of Congress to single-handedly block much needed promotions and medical care for scores of members of the US Armed Forces. In fact, he was referred to by The New Republic (Tori Otten. 12-5-23) as the stupidest senator. This is a sick type of government according to one communist male

---

[87] VICE News. "South Korea's Untouchable Families". 2022

[88] Nick Hanauer. TED. "The dirty secret of capitalism and a new way forward"

so it shouldn't be introduced to other countries. But realistically, it's the best one we have in this era.

From a related perspective, any positive social and economic progress of any country has to come **from within** its own people. Negative national experiences like poverty and oppression usually emanates from within the country itself but rarely caused by the acceptance of foreign deals and ideas from foreign entities. Another significant flaw of democracy in the US is the election of people with absolutely no experience in finance and economics handling the billions of government funds. In 2023, statistics from the Administrative Office of the US Courts confirmed that a former President's appeal has been awaiting a decision for almost **three times** as long as the typical D.C. Circuit Court – unequal treatment of plaintiffs.

Many US millionaires like MacKenzie Scott, however, who donated $275[89] million to women's healthcare provider Planned Parenthood in early 2022, was **not an avid Christian**, but **donated** millions of dollars to community projects while some of these astute preachers **demand** millions from their needy congregations. During this period, a Missouri pastor disavowed his congregation and stated that they were "broke, busted for not buying him a luxury Movado watch". Sorry preachers, but I'm warning you by using the WORD (Bible - your Guide) from YOUR BOSS (Jesus Christ). Listening to these millionaire televangelists is like listening to "sounding brass or tinkling cymbals" (1Corinthians13: 1). I can even debate[90] this concept with any of these fraudulent preachers if they visit our beautiful islands. I'm so confident that these preachers will agree with me (after our debate) that I am now ready to bestow him/her with a traditional Samoan chief title from my clan to celebrate the acknowledgment of their mistake. Hopefully they will remember the poor (Gal 2:10).

89

90

One proof of my view is an incident that happened in the Tomorrow International Ministries church in Brooklyn where armed robbers robbed the pastor during a service and escaped with more than $400,000 worth of jewelries that belonged to the pastor and his wife. Investigations later found the pastor has a criminal and deceitful past. If enriching pastors while thousands of ordinary people are starving is on the Christianity agenda then we should re-examine the scriptures more carefully. Again, people should do their own research of the scriptures and ask for help from many resources now available on the Web and not rely solely on pastors and preachers. I believe in Jesus Christ and the scriptures, but I don't trust many of the leaders of the churches who made the rules using their own personal interpretations of the scriptures. I have suggested to my children not to marry certain people, like the Hassidic Jews, since they discourage the use of the internet, and also people from religions that don't eat pork since this is one of the preferred foods of the Samoan chiefs, like myself. The main problem here is the wrong interpretations by church leaders.

Some of the restrictions demanded by some religious communities, for example the absurd ban on social media and running water! (Mennonites) would absolutely be ignored by the Samoan community. Yet, Samoans who are members of the Jehovah's Witness sometimes abide by their rules of not joining the military, not accepting blood transfusion, not voting in national elections, and the most annoying thing to some of my friends, is their refusal to support the national rugby team, Manu Samoa, the pride of our nation. Samoans are very proud and patriotic people, except this group. All this "mess" rooted in varying religious doctrines, seem to divide the world and not bring people together. Religion has and will continue to split humanity into a multitude of discrete entities which I think has contributed significantly to the global unrest and which political, racists and bigotry views have already triggered many global problems. Overseas, the most renowned atheist Christopher Hitchens said that the most dangerous religious group that has caused the loss of many lives is the Islamic militants. Because I don't know much about his statement, maybe the Islamists should visit him or his followers for an explanation. Chris also suggested that nations should adopt a secular constitution in order to live in peace. I think he gave Kosovo as an example.

A US right-wing televangelist who refused to get the COVID19 vaccine, begged viewers in 2021 to send him cash to purchase a private jet so he won't have to deal with vaccine mandates which he called "the mark of the beast". He already owned a fleet of private jets and also knew that scores of children world-wide die of hunger every day and yet he demands more private jets. I recommend to all Samoan people not to listen to this deceitful deranged joker.

According to HOUSTON (AP), Jan 14, 2021, a Texas megachurch pastor and former spiritual adviser to two U.S. presidents has been sentenced to six years in prison for bilking investors out of millions of dollars. Caldwell "used his status as the pastor of a mega-church to help convince the many victim investors that they were making a legitimate investment, but instead he took their hard-earned money from them and used it for his own personal gain," Acting U.S. Attorney Alexander Van Hook said in a statement.

In 2015, while millions of people, including children were starving all around the world and even in the US, God allegedly told one US televangelist to build a multimillion dollar 18 thousand sq.ft. mansion for his wife. (Houston Chronicle). In early 2024 a US Pastor created worthless cryptocurrency targeting Christians to support his lavish lifestyle, because "God told him". What kind of religion is this? Wrong question. It should be "What kind of evil man is this?. This is definitely not Christianity, and it's also happening in poor places like South Africa in churches like the 'Church of the Incredible" where the pastor owns several homes and about 30 cars. If this was truly a Christian establishment why is the pastor surrounded by armed men outside and inside the church, asked one Bible scholar from *Falealili*.

Another televangelist and his church in the US were ordered to pay $156,000 in restitution for selling a fake COVID cure (Cheryl. June 24, 2021). Pastor Bakker advertised a supplement called "Silver Solution" in 2020 that he falsely claimed could prevent COVID-19. It contained colloidal silver, which health authorities say is dangerous to one's health. The TV preacher and his guest Sherill Sellman claimed in the previous February on the Jim Bakker Show that the "Silver Solution," which comes in liquid bottles and gel tubes, could "kill" and "deactivate" the COVID virus within 12 hours. During the show, Bakker advertised the solution in a $125 starter kit. Per local Missouri news outlet Riverfront Times, Bakker's website also listed a case of 12 bottles for "$300 or more."

A female televangelist from St. Louis, one of the most popular female preachers in the world made *Time* magazine's list of the "25 Most Influential Evangelicals in America." The television ministry, *Enjoying Everyday Life* is dedicated to sharing God's love and the life-changing message of the Bible with the world through TV, radio, various media productions, [and] live conferences," according to the ministry's website. Her ministry provides her with a $10 million private jet, several homes that cost up to $2 million each, and a $107,000 silver Mercedes sedan, according to the *St. Louis Post-Dispatch*.

A Nigerian pastor and senior pastor at the Kingsway International Christian Centre (KICC) in London, reportedly earned £100,000 annually. He hosts the radio program *Winning Ways*, which is syndicated in London and Amsterdam. The television equivalent broadcasts in Nigeria, Ghana, and Zimbabwe on the Trinity Broadcasting Network, and in Europe on GOD TV and the Inspirational Network. The majority of his $millions comes from selling Christian literature and documentaries from his media company, Matthew Ashimolowo Media. Temitope Balogun Joshua, known as Pastor T.B. Joshua, founded The Synagogue Church of All Nations in Nigeria, where he runs renowned healing sessions (some of them exorcisms). The church has its own television channel, Emmanuel TV, which broadcasts Joshua's sermons and delivers his

alleged miracles to millions. He has a huge social media presence – his YouTube videos have been watched more than 240 million times.

The late Rev. Billy Graham died at the age of 99 in 2018. *The New York Times* named him the unofficial "national clergyman," and he preached for decades. Pastor Billy Graham was among the most influential Christian leaders of the 20th century. Graham was known for hosting the annual Billy Graham Crusades, from 1947-2005, with an estimated 3.2 million people who agreed to "accept Jesus Christ as their personal savior." His radio and TV broadcasts had a faithful audience of 2.2 million. Pastor Billy provided spiritual counsel for U.S presidents throughout his career, from Harry S. Truman to Barack Obama. He left behind a fortune of real estate holdings and book royalties, but I don't think he was a millionaire preacher. I personally like this preacher because we both believe that there are UFOs out there. I documented a related story which "proved" this belief in my previous book, which is a story conveyed to me by my late friend and priest "Tommy", who actually saw a flying saucer while playing cricket at Savaii about 40 years ago. Recently, scientists have discovered building blocks of life found floating around Milky Way, suggesting that "we are not alone" and that aliens have visited us in the past. Additionally, the US government during this period, unclassified many documents relating to UFOs and these have greatly attested to the existence of these "creatures". This televangelist didn't preach the twisted Prosperity Theology.

Recently, some Muslims, like Hamza Karamali, got interested in the phenomenon of UFOs and promoted the idea of Astro-biology and exo-theology; examinations of biology and theology pertaining to outer space. Interested readers may want to the video "UFO Leak of the Century -UFO Revers Engineering", it might make you a believer.

A Televangelist from Georgia, where he founded the non-denominational World Changers Church International. He helms the Creflo Dollar Ministerial Association, Creflo Dollar Ministries, and Arrow Records. He held his first sermon in an elementary school cafeteria in 1986. He preaches to a congregation of 30,000 members, and an estimated $69 million in revenue is brought in through cash collections. His prosperity, gospel-style teachings are manifested in his own lavish lifestyle. He owns two Rolls-Royces, a private jet, and three-multimillion-dollar homes.

A pastor, affectionately called Daddy GO, is from Nigeria. He is the general overseer of the Redeemed Christian Churches of God, where as a pastor he earns $2 million in revenue per year, according to *Forbes*. He also owns a gospel television station named Dove TV, which airs his sermons and other church programs. A $65 million Gulfstream G550 private jet and several houses are among his prized assets; and we usually thought that Africans are not rich people!.

Another example is a US televangelist, a *New York Times* best-selling author and the senior pastor of Lakewood Church, the largest Protestant church in America. Based in Houston, he founded Lakewood's television program and produced his father's televised sermons for 17 years until his death. His TV sermons have an audience of over 20 million monthly viewers. Sirius XM even had him guest host his own talk show. He focuses "more on the goodness of God and on living an obedient life rather than on sin." His biblical teachings emphasize the power of love and prosperity. Some of that can be seen in his $10.5 million home. Just half of his fortune can feed thousands of starving children all around the world.

Retired injudicious and misguided televangelist Pat Robertson resurfaced in early 2022 and made a controversial statement that President Putin of Russia was being compelled by God to not only attack Ukraine but use that country as staging ground for a war against Israel to fulfill biblical prophecy. What a madman. He passed away during this period and will be remembered by his ridiculous prophecies. This was a weird development because Israel was allegedly reported to be interested in acting as a negotiator between Russia and the rest of the world. I don't think there will be an attack on Israel by several nations given the close ties between Israel and some of the superpowers with nuclear arsenals. I think,

Armageddon should be interpretated in another manner, which should exclude a nuclear war. It's like the idea of building a new house of worship in Jerusalem before the second coming of Christ, where the new church building can be interpreted as the "new fertile grounds in the hearts of men getting ready for Christ's second coming". In September 2023 however, the Hamas group attacked Israel. There were about 20 Samoans living in Israel and another 20 were visiting when the Hamas invasion occurred. Armed groups should attack other armed men, like the Israel military, and not private citizens according to one farmer of *Falealili, w*ho also added "but then Hamas is a terrorist group made up of cowards who attack women and children".

On a similar subject, Pastor Franklin Graham (son of Billy Graham) announced that people should pray for Russian President Putin. After seeing and reading the world news about the Russian army killing hundreds of children, mothers, and patients inside hospitals, and also destroying churches, I concluded that these were absolutely the most shattering, demonized and demoralizing words that came out from his mouth and it was really damaging to my faith that I thought I would never, ever listen to this preacher's televised services again. Readers who disagree with me are very welcomed to talk to family members of thousands killed and displaced by the 2022 invasion of Ukraine during this period and also to the newscasters documenting this atrocious event. In fact, the Wrap (U Gonzalez 4/16/22) wrote "You (Pastor Franklin) **pervert the Word of God". I agree 100%.** Maybe we should still pray for Putin in 20 years from now, after praying for the innocent people killed because of the coward Hamas invasion. I don't think so.

This is the age where Christians should start studying the Bible on their own or join a study group where everyone has a chance to express their interpretation of the Word instead of listening to some of these sickening television and stadium services where televangelists (I sometimes refer to them as clowns and deliberate liars) that ask for money from people who are trying to make ends meet. An example of these people misleading the public can be illustrated by one preacher 2022 assertion that when he started to have grey hair, he spoke to his grey hair and it was gone (Justin Peters podcast). I hope he doesn't visit our islands, commented one local fisherman with grey hair. Furthermore, if these clowns can really heal other people, why can't they heal themselves first?. Why don't they just go to all the hospitals and heal everyone?. Apparently, he has a pacemaker!.

Various resources are also available on the internet including Hebrew and Greek translations[91] where interested people can study and get in-depth knowledge of the Bible **for themselves.** Research and self-study on various topics can sometimes acquire interesting and un-related fun information. For example, I came across the names of Mary's (mother of Jesus) midwife, Salome, and Santa Claus's wife – Martha. Questionable but interesting.

Televangelists live in expensive homes with assets worth millions and don't understand the realities of life experienced by most of their parishioners. Samoan communities therefore need to study the Bible for themselves as this will allow a more practical, realistic and more tangible application of Christianity to their local lives. Real, compassionate, Godly evangelists are those that actually live, occasionally experience, or come in contact with those living in poverty. Their sermons would be practical, relevant and truly biblical.

Ultimately the masses in overseas countries, will find out they have been lied to all these years by these preachers. This observation is proven by the fact that two of the most well-known preachers of the Prosperity Theology, Benny Hinn and Creflo Dollar later on had a change of heart regarding not only the Prosperity Gospel but also tithing. In fact, Dollar announced this change in his sermon entitled "The Great Misunderstanding" during which he apologized for his previous misguided preaching. Other preachers like Rev. Walter Wwambazi (TEDxLusada 10/18/20) had blamed religion for the poverty in Africa and also said that "religion is the number

---

91

one tool of mind control and that Jesus was actually anti-religion". According to Samoan Pastor Iliafi Esera, Jesus attended church but didn't attend a religion. According to a local man I interviewed for this book, in order to show genuine repentance, these millionaire preachers should return ALL the money they received from their church and maybe check themselves into the nearest Police Station or a Psychiatric hospital. I think this is too harsh though.

As a professional musician, the Prosperity Theology reminds me of a song titled "To obey is better than sacrifice, by Keith Green with the line "I don't need your money, I want your life". Interesting though, this song changed the life of one Dr. Paul Lim from an atheist to a Christian. Many years before these incidents, Martin Luther couldn't hold back his disagreement with the interpretation of the Bible during his time and fearlessly launched his theses, which was a "list of propositions for an academic disputation". This is similar to the very public objections by former HRPP leader and members of the National Council of Churches towards a misguided sermon by the SDIAC Church leader Pastor pau, during this period.

Modern day false preachers and cult leaders are **not to be solely blamed** for the after-effects of their false messages. Parishioners and the public who attend these services are also to be blamed. If people adhere to the proposition of self-study, fasting and praying, there will be very little chance they will be victims of these false Bible interpretations. In these islands, there are also instructors at the Malua Theological College and The Piula Theological College who may help the public interpret the Good Word. Visit them and be amazed how these local theologians can better interpret the Bible than the clowns mentioned above.

Unfortunately, it is sometimes hard to persuade people who have been brain-washed by false prophets, greedy preachers and cult leaders, and many followers will continue to give money and stay with these queasy actors. In early 2023, it was reported that the Church of Scientology leader David Miscavige was served 27 times with human trafficking lawsuit documents. Parishioners of this church should now know of this event and should immediately make informed decisions **themselves**. This problem is like the drug problems infesting these islands and the blame in not solely the suppliers **but the users themselves**. According to one of the chiefs in my village, if there are no users then there will be no suppliers, and therefore no drug problem. Simple economic principle from his taro plantation business.

This is the age for this phenomenon of self-study to happen and this is the most important message from this author. This is "financially" benefit since it will cut off the middleman and the individual will pocket the whole profit – the truth. The increase in the number of Samoan Bible Scholars and pastors who have been well educated in foreign seminaries, Bible Colleges and universities on the Word of God, and have utilized various Bible-related resources on the internet is a positive start. Remember the Holy Spirit is available 24/7, and for the elderlies who are not familiar with new technologies, please utilize the young generation, especially your children and grandchildren. Most of you have paid for their education and maybe for their laptops and iPads, now it's time to utilized their contemporary knowledge to assist you understand the truth about The Gospel of Jesus Christ. It is much better to falter after self-research and praying than being led astray after blindly following someone else's sick views.

There are also several false prophets and pastors in the Samoan Islands. A few had preached the end of the world during the 1999-2000 Y2K internet era. Their preaching resulted in the gathering of many people in front of the main government building in Apia, waiting for the "end of the world" as the calendar year went from 1999 to 2000[92]. I had a good sleep that night as I had taken care of the American Samoa Government (ASG), Marine and Wildlife Resources (DMWR) computer systems, weeks before the "end of the world", as DMWR's IT Supervisor. Additionally, a few local pastors

---

92

have continuously predicted the second coming of Christ for many years. One writer wrote: "a little knowledge is very dangerous", and may I add "it's also very demeaning to the false pastor's family and his church". I personally dislike these people because there was never a public apology and they continue to live arrogant and unrepenting lives as though they didn't do anything wrong. Like overseas megachurches, locals continue to be members of these false pastors' churches and one very physically sick one[93] has built a very huge and expensive church building in downtown Apia during this period. The public in American Samoa has been continuously reminded of Jesus's potential second coming through certain individuals and some groups showing billboards along the main road for several years. This should stop and alternatively, people should study the Bible, fast and pray.

Other types of unfitting pastors exist in these islands. An arrogant pastor decided to build a home for himself on communal land without giving notice to the Chief of the family in the village of *Fagatogo*. Near the same area, in 2023, a Methodist pastor threatened members of one big family with a Samoan cricket bat (I interviewed two chiefs from that village regarding this incident).[94] As a result, one of the chiefs took a break and decided to wait until a new pastor is conferred before he returns to the church. A similar event occurred in this same church a few years ago with a former pastor. A former pastor of one of *Alataua* district's American Samoa CCCAS church was asked to resign after a very prolonged period of medical treatments overseas. For many months, the church faithfully kept sending lots of money for the sick pastor. Upon his return and consequent resignation, the egotistic man decided to get even and exhumed the body of his son who was buried in the church compound despite the consequent serious stench that overwhelmed the poor people that assisted with removing the body.

93

94

A couple of pastors I interviewed refrained from condemning these unruly church minsters but agreed that their actions were definitely unwarranted and not biblical. In Samoa, Court Judge Viane-Papalii sentenced Rev. Poasa Leaupepe to six months probation for assaulting a 16-year-old who died of leukemia shortly after he was allegedly assaulted by the pastor. The Court also ordered that the pastor serve sixty hours of community service. The pastor asked for a discharge without conviction but was denied.

During this period, a EFKS pastor of a village near Apia refused to offer assistance to the Faumuina family, who were banished from that village, even though members of this family were assaulted and their family home was damaged by parishioners of his church, under the directions of the chiefs. In these islands, writing or speaking out against pastors is generally culturally taboo. If I am called upon to be a pastor of this village I would refuse because of safety reasons!. This reminds me of the story about the Good Samaritan and would continue to write about "taboo" issues. Scores of humble pastors who have graduated from Piula, Malua and several other local and overseas Bible Colleges are waiting to be called to serve Christ. Let these **humble** servants of Christ serve these villages please, and release the arrogant ones. They were not fit to preach the Gospel in the first place. Readers are welcomed to discuss this issue with me and be reminded to bring your Greek-Hebrew Bibles.

Samoans experience false faith healers during Benny Hinn's crusade in Samoa in 2001 where **no one was healed**[95]. A family member who used to attend local AOG church services when visiting pastors would preach and invite parishioners who need healing to come to the stage so they could receive the healing power of God, after they touched them or just waived their hands towards them, admitted to me several times that she was not healed and would not fall down (as seen on TV) as other people onstage when attending many of these types of services. But readers should keep an open

---

[95]

mind and ask for proof. My I suggest to take these Faith Healers to Saint Jude Hospital for Children and other similar institutions and have these fake people do their thing. I know you will not be disappointed with the results and their excuses. Please let know about your experiences. I know all of the sick children will promptly walk out of these hospitals completely healed. Absolutely not.

In early 2023, I visited a friend, in Apia and he was with some friends and a pastor. I asked him later what the good pastor was doing in his garage and my friend said that he was looking for some dope to smoke. I have to include this event in this book to illustrate how a **very few** bad apples contaminates the great work done by the churches and the scores of faithful pastors, deacons and priests around these islands. Doubters are welcomed to drop by my home in *Malaloa* to discuss or debate this true and real issue.

I'm not the only one who is very disgusted with these deceiving and phony preachers on television, there are scores -if not hundreds - of appalled people out there including Pastor John MacArthur[96] who wrote:

Someone needs to say this plainly:

> *"The faith healers and health-and wealth preachers who dominate religious television are shameless frauds. Their message is not the true Gospel of Jesus Christ. There is nothing spiritual or miraculous about their on-stage chicanery. It is all a devious ruse designed to take advantage of desperate people. They are not Godly ministers but greedy imposters who corrupt the Word of God for money's sake. They are not real pastors who shepherded the flock of God but hirelings whose only design is to fleece the sheep. Their love of money is glaringly obvious in what they say as*

96

well as how they live. They claim to possess great spiritual power, but in reality, they are rank materialists and enemies of everything holy".[97]

In 2007, Senator Chuck Grassley opened a probe into the finances of six US televangelists who preached the Prosperity Theology and release his review of the six ministries on January 6, 2011. There are, however several genuine televangelists preaching on US national television – bless their fine and genuine souls.

The Dead Sea takes in water but doesn't release any water, and this body of water will slowly die. In fact, its waters have significantly receded in the past decades. This concept is like these faith healers, deceitful so-called prophets and deceptive evangelists who advocate the Prosperity Gospel. *They take in the poor people's money then refuse to give it back, and then they will slowly die out in the next few decades.* This group of pastors have some convictions parallel to atheists' beliefs.

There's a group of televangelists that believe sick and impaired people are not healed because they don't have enough faith. This is a lie they give when they couldn't heal sick people. Readers can get a realistic and true perspective on this issue by watching testimonies by a blind preacher name Kayumba of Zambia, Africa.

To all the Samoans reading this book, please know that God never spoke to or anointed the false preachers mentioned above and also please, instead of watching these deceitful preachers on television, watch Manu Samoa rugby games; wrestling matches (featuring our own The Rock); David Tua or Joseph Parker's boxing matches or better still, watch podcasts and internet presentations by Justin Peters[98] John McArthur, Jay Smith, Jason Lyle and other Christian professors instead. I promise you; you will be immensely blessed.

97

98

On a related matter, the cult group Shicheonji, founded in 1984 by Lee Man Hee, started recruiting for new members in Samoa during this period. This group is also known as Mount Zion, and is from South Korea. In Fiji, it called itself World Peace, Heavenly Culture and Restoration Light. Former members of this group have expressed concern because of the "spiritual abuse, depression, post-traumatic stress and damaged relationship with families" from being involved with this group (Stuff New Zealand).

To be "fair" to these preachers, I should mentioned that there are other types of people like scammers and motivational speakers[99] who have extorted millions of dollars from the public.

Readers who find this topic interesting (or sickening) may also watch *They got caught then this happens- John MacArthur The Gospel of Christ*. 2023 and the video *When scamming your parishioners goes wrong*. I assure the readers that after watching these presentations, you would agree with my analysis above. If anyone is still in doubt of the descriptions of these false and deceitful preachers, well, Dr. Justin Peters podcast refer to these people as "heartless, uncaring, wretched, narcissist". Amen from these islands.

Before I end this section, I need to provide some advice to these Prosperity and Faith healing televangelists (not the scores of genuine spirit-filled ones out there): please go preach in Syria, Pakistan and other countries in the Middle East; the slums of Philippines and India, Africa, China and Utopia in Australia. To really test your faith and confidence in what you preach, why not join and support the Islam expert Jay Smith at public debates where there is always danger of being killed by Muslims in the streets. There is this Samoan saying: "*E iloa le tautai i aso afa*", meaning a: boat captain's capabilities can only be gaged during rough weather. God needs you in these places, not in arenas in the US where you are comfortable and inundated with numerous naïve people. It will also be very easy to do this since you have many private jets and lots of money. I know what I'm talking about. My parents, when they were serving as LMS missionaries in Papua New Guinea had to walk for **nearly a week** from Port Moresby through the jungles of that country to their assigned place of work.

Theres a difference between getting rich from one's hard work and getting rich from money donated by the poor and desperate sick people yearning for healing. Amen.

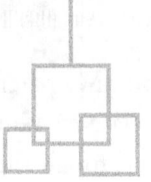

Commentaries - Samoan Style

"Challenged" Samoan people that are seen begging in public places **do have loving families that welcome these people to their homes anytime.** As proof, one can't observe any of these people sleeping on the streets at night, they're sleeping at their homes. Family members (and some NGO's and government agencies) occasionally go by and try to help these poor souls but their mental states often thwart these efforts.

Secondly, television overseas news often show people being harassed, abused and attacked in public places (like subways) while onlookers refuse to assist the victims. The increase use of cellphones to record any event has helped with the increase of this "onlookers" phenomenon. I previously believed that in these islands, there are absolutely no onlookers. Any Samoan in the vicinity of an attack will surely assist the victim, and if these types of events happen overseas to anyone and there are Samoans around, they will certainly try to help the victim. I was partially wrong. In around August 2022, two National University of Samoa (NUS) female students had an altercation for several minutes while at least seven students just watched without lifting a finger. A video of the fight went viral. Anyway, the altercation seems to be a draw and my theory applies only to public places.

As a Samoan Preacher's son, I have always wondered why I grew up in a very poor family. I always thought that if my parents were serving the Lord, I would have a comfortable life, but it was just the opposite. I sometimes walk to school (Samoa College) without footwear and most of the time without lunch or breakfast. My father's relatives, especially Aunt Faaliliu and Ainofo, would always come by to give my father some money to help pay for me and my siblings' school and bus fares for schools in Apia. Recently, I began to understand that Jesus chose single men to be his disciples and not those with wives and children. His command "no bag for the journey or extra shirt or sandals or staff, for the worker is worth his keep" corresponds to single men-disciples lives and not disciples with wives and children. Maybe if I had prayed for a better life I would have been blessed, but I didn't. Furthermore, the answer to such prayer might not be positive (see my reflections on how I think God answers prayers in another section). But life is so complex and mysterious and I might have gotten what I had hoped for but with a different outcome. For example, who were praying earnestly to have a child didn't get one but were blessed with two. One of them turned out to be a murderer. But for me, I was very fortunate, because it was through those trying times that I learned (mainly struggled) a lot about life, and partially sealed my partial belief in destiny which eventually brought me to write three books about church and religion. When Steve Jobs, the Apple and iPhone former CEO, finally located his birth parents (he was adopted), they didn't know how Steve would turn out. Steve's destiny, however, is known throughout the world. Steve felt very sad and disappointed when he found out that he was adopted. However, his adopted parents assured him that he was chosen. Destiny can be very odd sometimes.

I would like to suggest that the two Samoa judiciary systems should jail the foolish locals who have been predicting the second coming of Jesus several times for many years and have swindle the true Christian resolve of many people. If Jesus himself doesn't know when he returns then how in the world do these liars know it. These Samoan false prophets should at least offer a public apology. They haven't and never will. Pope Francis apologized to the Canadians in regards to a Catholic-affiliated residential schools issue and Dr. Puni of the local SDA church explained that their elders erroneously predicted the second coming of Jesus in the past, so why do these liars refuse to apologize. Like the Lands and Titles Courts, a Church Affairs Court should be established to jail these Samoan false prophets so that our social environment is free from their nuisance and also let their families know of their stupid and thoughtless deceit. This is not a new idea. It will be like the EPA's (Environmental Protection Agency) mission "to clean up the environment". It is also not farfetched. Consider Iran's Morality Police Force which enforces the rules of their Muslim beliefs, which in 2022, resulted in imprisoning a female for disregarding the rule of wearing a hijab; sentencing a woman to death in 2023 for adultery (AP. Nov.2, 2023); and executing 9 women because they were driving cars and some men without beads.

According to some researchers, the usual question about Noah's Ark not having the capacity to hold all the animals is because, according to one scholar, only the land animals were meant to be taken, and the related passage relates to "family" not the "species" (ask your High School students for the difference, this is taught in every High School Biology class). Some scholars have also noted that the flood was limited to inside the Old Testament world, not the globe. There are also new discoveries that seem to prove The Flood, like the evidence provided by Geologist Steve Austre Ph.D. There are also strong arguments that God actually commanded organism to reproduce according to Dr. Kurt (The Arphaxadian Epoch. 2022). But to be fair, there are tons of written materials and presentations against the story of Noah despite various fine analysis by some religious groups like Jehovah's Witnesses. For doubters, please examine the presentation by Dr. Major Coleman, Assistant Law Professor from State University of New York (Anil Kanda. "Renown Professor Debunks Evolution: Unveiling the Truth.").

with its magnificent replica and explanations of the Noah phenomenon. It will surely make anyone a believer. The interesting asset of Dr. Coleman's explanation is that he has offered any atheist $20k if they debate him on this subject in public. No one has come forward in the past 20 years!. This debate will continue for many more years.

Since several so-called scholars have stated that Jesus was married to Mary Magdalene, here is a very simple opposing view. When the disciples supposedly asked Jesus "why do you love her more than all of us?. Well, if Jesus was actually married to Mary of Magdalene he would have replied: "for God sakes I love her more because she's my wife you dumb nuts". But of course, Jesus didn't reply in that manner because He was NOT MARRIED to Mary Magdalene!. Very simple logic. If anyone asked me the same question regarding my relationship to my wife, I would have replied "for God sakes because she's my wife you idiot".

Some overseas wars in the past were fought because of divergent religious beliefs. Not the church itself but the interpretation of various convictions. For example, the war between the Catholics, Orthodox and Muslims in Bosnia killed many people and went on for an extended period, all in the name of religion. According to one of the chiefs in our village, there is a solution to this mess, and that is for all the people of those countries to accept (agree to disagree) and respect other people who have different beliefs and religions. This concept would have undoubtedly saved thousands of lives in the Balkans conflicts. He said this is a tough and complex solution but it's the **only one**. Rome wasn't built in one day. For the varying religions emanating from different countries, there should be an understanding that the God of the Jews became the God of Christianity and then the God of Islam.

Modern technology and the availability of several resources on the internet would gradually shine some light on various questions from atheists. This author, however can offer a simple answer/ explanation to almost of those questions, and that is: **God has His own divine plans** and some of the details may have been lost during the various translations of the Bible and consequently ended up reflecting mysteries. The Godly truth in the scriptures, however, never falters. Some of these queries, questions and uncertainties include:

- Abraham was born in a city allegedly ruled by people who supposedly didn't arrive about one thousand years later.
- Differences in the stories of the four Gospels. This is the easiest question to answer but I prefer that interested readers personally join me for a steak and lobster dinner (I'll pay) and would marvel at the simplicity of the answer.
- There is a **possibility** that the famous Census that brought Mary and Joseph to Bethlehem never happened.
- Jesus eats the Passover meal (Thursday night) and is crucified the following morning vs Jesus does not eat Passover meal but is crucified on the day before the Passover meal was eaten.
- More than one version of the Commandments and whether the term "Ten Commandments" refers to what is written in Exodus 34, Deuteronomy 5 or Exodus 20.
- How did Jonah survive inside the whale
- The two Creation stories in Genesis with different Orders
- The Flood seem to happen in two different ways
- Some scientists believe that the walls of Jericho were toppled long before Joshua's army arrived.

In the Samoan culture, the *tulafale* (Orators), always speak on behalf of the High Chiefs (like myself). This is very similar to the role of Aaron, who was chosen to speak on many matters on behalf of Moses (Exodus 4:14). Remember Moses told God that he doesn't know how to speak in front of Pharoah and God said to him: "Now therefore go, and I will be with thy mouth and teach thee what thou shalt say" (Exodus 4:12). Samoan author Tuiatua Efi eloquently wrote regarding this concept: "*E fa'agaganaina oe e le Atua Fetalai*".

Science can surely collaborate with Christianity to bring about healthy conversations regarding various aspects of life. Oxford Mathematics Professor John Lennox (Veritas Forum 2012). said that atheism is actually faith-based and that faith (not religion in general) is essential to science. I agree with the professor after years of research for this book. According to science: energy can't be created nor destroyed. The Christian God had already created energy in Genesis. HE breathed energy into Adam and Eve and our souls (a form of energy) will be forever be around, like my grandmother Sina who visited me 40 years after she passed away[100]. People of Vanuatu believe that their dead are still alive but in another spiritual dimension. (South Seas Island. Nadine Amadio. Augus & Roberston. 1993).

The passage in the Bible in Mathew 18:20, 'for where two or three gather in my name, there am I with them", was explained by one scientist using a law in physics called 'constructive interference", which states that "when two wavelengths of equal lengths meet, the size of the frequency doubles". He noted that when two or more waves meet and are in-phase, they combine to form a wave of even greater amplitude. According to this research, the energy for this phenomenon comes from the part of the brain call gray matter. Is this where the souls resides?.

Ponder this, the hospital operating room may have been believed to be the most sanitary place on Earth, but it's not. It's the production areas where microchips are manufactured!. The largest organ of the human body is not inside the body; it's the skin, or some people believe it's the endothelium (layer of cells lining the blood vessels). The World's largest plant is not a "single" plant but a vast seagrass meadow in Australia[101]. According to some aviators, the airplanes' Black Box often retrieved by investigators after a crash is actually two boxes and these are orange, not black. Identical twins don't have the same fingerprints so they are not identical. The General

100

101

Sherman Tree is the largest in the world – not the *mosooi* tree - by volume, at 52, 508 feet(1,487 cubic meters) and towers 275 feet high and has a circumference of 103 feet at ground level. Peanuts are technically legumes not nuts. Mary Magdalene is never referred to as a prostitute in the Bible – as some scholars[102] have alluded to, and Jesus has about seventy-two disciples (which included some women)– compared to just twelve Apostles. Peaceful protests didn't start with Martin Luther King Jr. in the US. According to John Dominic Grossan, it started in the first century Jewish community. In these islands it started in Samoa during the *Mau* Movement in the early 1900s and culminated in December 28, 1929 (Black Sunday) with the assassination of Tupua Tamasese Lealofi III. - Martin Luther King Jr. on the other hand, was assassinated in April 4, 1968). Not all giraffes have normal spots. A few, without spots can be found in Japan, Uganda and Tennessee. Dementia doesn't happen to only old people, it also occurs in young people and it's called "childhood dementia" a genetic-related condition. Contradictions pops up everywhere, even in religion.

A few years ago, a researcher claimed that Jesus had brown skin and members of the Catholic Church believed that the people who initially saw the Lady of Quadalupe attested that she has dark skin. In more recent scientific research (British Researcher Richard Nealy) it was recommended that Jesus is most likely a dark-skinned man. Televangelist Jesse Duplantis had a vision of Christ during this decade and he said on US National Television that Jesus had brown skin. I think this evangelist was under the influence of drugs because he later said he had an encounter with a horse in heaven, beside other ridiculous revelation from his heavenly experience (see web contents refuting these claims). Africans consists of not only white and black-skinned people, but there are also brown-skinned people called Bushmen who are the original inhabitants of South Africa. Samoans are brown skinned people. Maybe Samoans should look

---

102

into their DNA, maybe they'll find some connection to Jesus?. Ok, don't be upset it's just a little satire to wake you (reader) up. I also include this Comic Relief because I believe that the smartest people on this planet are the professional standup comedians.

Samoans have assisted some families with their personal issues in dangerous places like Italy where the mafia is well established. A well-known New Zealand journalist, Mr. Lomas assisted a Samoan-Tokelau family locate a relative who had been living in Rome for thirty years. The help of Catholic parishioner and Samoan Chief Papalii Giovani C, a Samoan living in Rome, who is also the Honorary Samoan Consul for Samoa, assisted the Filipo family locate their relative and the lost relative was finally located in this dangerous location.

Several Samoans, including Bible College students and instructors, pastors and interested parishioners keep up with world news regarding religion in overseas countries. The existence and improvements to internet systems and other resources available in these islands had made it easy to read about and study these events in relation to the Bible. This has allowed the locals to evaluate and assess related events instead of relying on foreign missionaries and pastors to interpret the Bible for them. Consequently, a more relevant approach to teaching and understanding Christianity emerged in the past two decades and this included the subjects discussed above. More locals have already been misinformed in the area of religion and politics as the number of people who have access to social media and the internet increased rapidly in these islands. Talking with some of my friends, associates and several other people in my community, I have noticed the unfortunate impact of internet misinformation regarding religion within these islands. Add Artificial Intelligence (AI) to this mix and we have a soup of serious magnitude. Sometimes, the lack of information available on the internet is just as unjust as misinformation available on this platform. Take for example the momentous Tank Man event that happened at Tiananmen Square in China in 1989. If one does a

Google search of "Tank Man" in the US, several pictures of student protesters and military personnel firing at the crowd would appear. Do the same search within China and happy faces of tourists smiling will appear, **minus the protest**. When a few students from Beijing University were shown the famous photo of the Tank Man during this period, none had any idea of this major political event. There goes the integrity of the university's History and Political Science Departments and I hope no Samoan student should ever enroll in this deficient program.

The approach of self-study, will lessen the probability of innocent people joining cults that have killed and murdered their own members in the past decades. Definitely not in these islands but in overseas nations. Even world-renowned atheist Christopher Crtichens advocated self-study. Mark 14:62, noted that Jesus will be coming "in the clouds": People planning to join a cult should first ask themselves if their cult leader who usually claim to be the Christ, did come in the clouds. If not, then just dump the group or seek more information and advice before being duped. As simple as that. Additionally, self-study and research (in any field) would also lessen the negative impacts of misinformation. In the US during 2022, a few documented cases of some smart adults who succumbed to the false Qanon creed consequently killed their relatives. The US Supreme Court had certified the national elections but still, misinformation continued to misadvise the public, facilitated by supremacist ideologies which are clearly misguided and non-Christian concepts. Many were eventually jailed.

Prolific Public Speakers like Minister Farrakhan had offered ridiculous and impractical solutions to problems like segregation where he suggested that all black prisoners in the US should be released so they can return to Africa to build a Nation of Islam or provide reparations in terms of money and land. This will NEVER happen on this Earth. If people of his faith do their own research and think for themselves, then no one will attend his campaigns. Furthermore, Malcom X, another expert on Islam, provided a very different take on related issues which further confuse the followers of Islam. From another perspective, a 60 Minutes Documentary posted this statistic: About 74% of extremist killings in the US in the last 10 years have been carried out by right-wing extremists, not Islamist

In the Samoan Islands, different denominations (estimated around 35[103]) continue to proclaim various doctrines during this period. As I wrote in my previous book, many unsaved souls like me would appreciate a single settled Biblical doctrine instead of the diverse ones confusing us and declared by confused people. A good example is the conviction by AOG that its parishioners shouldn't perform the Samoan *siva* (traditional dance) but then watching Cardinal Pio Taofinuu, the most prominent Catholic clergy in the pacific for the past fifty years, performing the *siva* on KVZK TV, created reservations regarding this prohibition by my church. I said to myself, if my church is right, then I don't think the poor Cardinal will have any chance of getting through to everlasting life!. More *siva* performances by members of the AOG church were seen during the Samoan Heritage Week celebrations in Hawaii during this period. One that stood out was by a Samoan AOG pastor who used to be a staunch supporter of this prohibition and was also wearing a Samoan tattoo. Very weird people. The Mormon Church used to be against LBGT members holding hands, kissing and other similar acts. Years later the Church partially allowed it, then later disallowed it (Fifth Estate 2023). For related information, readers can examine *The Mormon Church in Canada. Where did more than $miilion go?")* or watch "Mormon Leak" website. Religion is confusing, so do your homework.

A single church and doctrine will undoubtedly bring harmony to the Samoan religious community. Yale theologian Newman Smythnoted said that religion has lost authority in family, community and intellectual life and a Christianity divided in its own house against itself could not survive. The irony is, Christianity may

---

[103]

survive in these islands if churches agree on a universal approach and doctrine but this will **never happen**! About five hundred years ago the German monk Martin Luther initiated the Protestant Reformation, a split in Christianity. The conventional wisdom then was that the Catholic church was the only entity with power and right to interpret the Bible. Readers should note that, before the Reformation, there were two main contributors to the idea of reformation and these were England Professor John Witclife (14th. Century) and John Huss. This resulted in deep divisions between Catholics and Protestants and contributed to violence and wars. In the 1900s, some protestants in the US started an effort to negotiate the "reunion of Christianity". They failed. A concept called "Christian Unity" with theologian Smyth, emerged and failed. The Pew Research Center recently put out a report indicating that the number of self-identified Christians has declined 12 percent points since 2012, while those that were registered as "nones" grew 10 points during the same period. This declined had been trending downward since Pew first surveyed this issue in 2007. The Bahais continue to believe in this unity notion. Pope Francis also seem to advocate this concept during this period when he invited Christians of all denominations to gather in St. Peter's Square to help further the cause of Christian unity[104]. There is also an annual week of prayer for Christian unity that ran from January 18-25 presided by the Pope in a Rome Basilica.

I have a suggestion, forget about churches and concentrate on the Gospel of Jesus Christ, which has concepts of unity for everyone. In contrast, I personally predict that Jesus will return and would find his people (*Amo Israle*) still fighting the Palestinians, which is another sign against the theory of Christian unity. To further illustrate my point, consider this narrative: An African went up to God and ask God when Africans will be free from starvation and God said, 50 years after you die, and the man cried. An Afghanistan man went up to God and asked, when will be my country men stop

---

[104]

fighting against each other, and God said, 20 years after you die. The man cried. Finally, a Palestinian went to God and asked, when will my country be free, then GOD CRIED.

I couldn't find good examples of humor in the Bible. But a contemporary event shows that it has its place in church-related events. In June 2022, the F-word made its debut on T-shirts at a papal general audience and a laughing Pope Francis gave members of the choir wearing them, VIP treatment (Reuters. Pullela P. 6/11/22). Apparently in the last two decades, humor has found its way into the sermons of some local pastors. A short trivial laughter can occasionally be heard in some congregational churches as some pastors describes some anecdotal event that may(or may not) help energize a boring church service or maybe enliven those with hangovers. This has never happened in the Congregational and Catholic churches during the first 50 years of Christianity in these islands, where the services are normally somber, eerie and lugubrious.

According to some preachers, God has plans for each individual. So, if one wants to see God laugh, one can just tell HIM your plans and one will surely hear a whopping laugh that will echo around the universe. Now is this humor enough!. I've just come across a similar story about an old lady who was invited to offer an invocation. After she provided the prayer, she was about to leave the pulpit but abruptly returned and said "God, sorry I forgot to ask you to please let me find my car at the parking lot because after the service last week my husband stopped outside the church to speak with some church members while I continue towards our car. There was a man behind the wheel who asked me if I was going home with him. I looked up and said I'm sorry Archbishop. Life can also be much better if people can add some humor to the usual tragedies of everyday life.

A man from *Toamua* Samoa pleaded guilty to one charge of bigamy in the Samoa Supreme Court in 2022. He was convicted and ordered to pay $100 for court costs. King Solomon and others in the Bible had multiple wives, so why are some believers pointing fingers at other churches who allow polygamy?. Ask your pastor. This is also an ordinary matter in other countries. Recently in the village of Bugisa in the Butraleja district Uganda, Musa Hasahya Kasera was reported[105] to have 12 wives, 102 children and 578 grandchildren. Scholars have also noted some related events in the world of Islam.

Bird nests from Malaysia are expensive delicacy in China and the delicacy sea worm *palolo* is **very expensive** in these islands. Some south pacific islands like Palau, have lots of *palolo* but they don't eat these worms. I'm planning to visit these islands and harvests as much *palolo* as I can in the future. One local fisherman, Elvin Mokoma, netted at least $10K for one night's fishing for *palolo* several years ago and he normally give some for all the pastors in his village of Pago Pago. Mr. Mokoma can be compared to the lady in the Bible that spread expensive oil on Jesus's feet (Luke 7:36).

Recent discoveries suggests that early sets of mummies were not made by Egyptians but by the Chinchorro of the Atacama Desert South America, thousands of years before. Again, new discoveries keep emerging and will continue to make interesting debates in the secular and religion realms. For example, readers may be interested in a book by Joel B Kramer called "Where God Came Down" which provides the story of the discovery of scripture scrolls dated earlier than the Dead Sea Scrolls.

The Russian Orthodox cross has two extra crossbeams. The top beam is where the sign "Jesus of Nazareth, King of the Jews" was placed. The second is where Christ's arms were and the bottom one represents Christ's footrest. In these islands, one would observe different depictions of Jesus's face in various churches and even a sculpture of Jesus with a Samoan tattoo. I would be very offended if someone put up someone else's photo/painting with my name on it. You figure this out.

Millions of people in overseas countries live in misery and poverty while the inhabitants of these islands mostly live relatively in peace and generally in satisfactory conditions. Documentaries found on the internet show young girls in Asia being sold in the main streets (some sold by their own parents) and current information shows that human trafficking has been increasing in significant proportions throughout the world (mainly because of the involvement of some crooked government officials and powerful organized groups according to NGO researches). Statistics provided by NGOs like Chab Dai Coalition shows that 80% - 90% of mothers sell their daughters in some Asian countries. Hundreds of young girls are forced into prostitution in some countries **because of poverty** according to Agabe International Missions. Many young men in some parts of India depend on stealing to survive. On camera, these young children proclaimed "if we don't steal, we will starve and die". Many young girls in countries like the Philippines and Jamaica turn to prostitute to feed their children. Most of these girls have no education and no work experience. Should Christians condemn these young mothers?. If you're a pastor and you condemn these people, I wouldn't dare join your church.

In Senegal, there is an outrageous program where young children are lured in to study the Quar'an, but in reality, these poor children are given small bowls every morning and told to go out in the streets and beg. The children are punished if they fail to get a certain amount of money (francs).

As thousands of immigrants illegally cross over to the U.S. in 2023, those from Venezuela stated on camera that "if they stay in their country, they will surely die". Lady Liberty used to smile down on these refugees in the past but now the economics of the migrant issue is overwhelming and the reality of its social effects may have dimmed her smile. How about having each country erect a similar Lady and work towards self reliance?. Its interesting to note that the culprits cultivating these miseries in poor countries are the refugee's uncles, fathers and grandfathers!. The churches can and should help with these conditions.

Would these thieves and prostitutes be forbidden from entering heaven because of their lifestyles?. Maybe this is why the Christian God is irrelevant to these millions of ill-fated people. Maybe there is a related reason why Christ forgave the women in John 8:11. Maybe I should stop assessing Biblical concepts altogether because I'm now sure I'm (and anyone else for that matter) **not getting anywhere**?. After much research for several years; utilizing my own 50+ years of experience with church matters and writing two books on religion in the Samoan Islands, I'm still far from understanding life in relation to Christianity but comparatively nearer to the truth compared to scientists trying to explain the origin of life!.

In relation to this, I respect anyone else's perspective on religious matters but would not agree to their convictions. I now have my own (e.g. personally experiencing *Telesa*). This is in contrast to Dawkins who utilize other scientist's data and experiences. I use the word "respect" because I believe that respecting other people's opinions and **agreeing to disagree** will bring much needed harmony to society. Rev. Billy Graham even though he has some doubts towards other people's beliefs, respected Muslims, Catholics and other people during his ministry. Respecting doesn't mean agreeing. A fabricator and rookie Kevin Hunt later said that Graham should not be friends with these people. This is a good example of jealousy, the lack of maturity and the need to promote respect for the broader good of humanity. Kevin is welcomed to stick his head into the *Pulotu* cave in *Falealupo* to get some insights into the afterlife and some religious insights?

Foreign countries experiencing continuous national conflicts should employ this simple and not so novel concept from these isolated islands in the pacific. During the Hamas conflict with Israel in 2023, there was already a 10-year-old organization called the Sisterhood of Salaam Shalom whose goal was to build trust and mutual respect between Muslim and Jewish women. In the Pacific, the coming together of different island people to promote their traditional music was authenticated in the establishment of the Signature Choir and NZ Symphony Orchestra – *Mana Moana*, which to my humble opinion as a professional musician, provides the very best from all the participating island nations.

A couple of atheist-scientists commented on national television that the Christian God didn't do a good job when he created man. He pointed to some physiological problems with the human body. I'm hundred percent sure that this scientist didn't see the final performances on America (AGT) and Britain Got Talent televised programs. This scientist would be baffled by the performances which usually highlights the magnificent craftmanship that went into the creation of the human body and mind. A chief in my village (also a deacon) told me that if he was watching these television programs with this scientist, he would tell the scientist to "eat my shorts". I'm not sure what he meant though and I can't ask him because he has since died. Can anyone possibly compare Miss Universe (created by the Christian God) to a robot?. Would any man sleep with a metallic robot instead of his beautiful wife?. Ponder this statement by Ernest Becker, a Pulitzer winner: "Man cannot endure his own littleness unless he can translate it into meaningful on the largest possible level".

Around 1980, archeologists digging around Jerusalem announced that they have discovered the tomb of Jesus. This exhaustive project attracted several critics doubting the discovery but the overall assessments seem to favor the notion that the coffins discovered contains the remains of Mary, Jesus and other relatives. A Christian Professor was asked via a documentary regarding this finding if this will change his views about Christianity and he said no. To me, this is a simple question. As I mentioned in this book, Christianity is purely a spiritual phenomenon and therefore the resurrection refers to Jesus's Spirit leaving Earth, maybe not the body, since Christ also has both a human and Godly dimension. The discovery doesn't change my perspective about Christianity. Additionally, the discovery needs to be scientifically assessed independently preferably by several other scientists.

One scientist turned Christian said during an interview that one of the things that delayed his attending churches was that he saw on television pastors yelling at the parishioners for over half-an hour during the sermons. A friend of mine, a former US JAG Judge told me he has the same reason for not attending church services. I suggest Samoans continue to attend church and if they're offended by these types of noisy practices, another church is right down the corner or the next village or maybe just don't attend a Pentecostal church service but watch church services on television (not those advocating the Prosperity Gospel). Don't let these petty events deter you from learning about the Gospel of Jesus Christ and spiritually praising God.

I've always wondered what would happen to people who have not heard about Jesus Christ and those who have heard but didn't have the opportunity or were unable to study Christianity due to various circumstances, like those under the Taliban, Communist and other anti-Christian entities. For example, the millions of people in Africa, China and various religious groups in the Middle East. Here are my two cents, which maybe from a **legal** standpoint. Jesus Christ is "the way" and he understands all the circumstances and I know he will be at the door and gladly open it for "**special circumstances**" cases. Special circumstances are like VIPs from the United States visiting these islands and don't have to go through Immigration check points. This concern about people (or aliens) not being able to be saved because they haven't heard about Jesus was recently entertained by a few Muslims in their study of exo-theology and Astro-biology (Hamza Karamali).

The name Jesus should be clarified by all pastors in all Samoan Christian services. It should be **Jesus Christ, not just Jesus**, because one can find scores of people name Jesus in prisons in some countries. A lawyer name Jesus Suarez was also one of a former US President lawyers during his 2023 controversial trials. Some men named Jesus have been jailed for using drugs in the US and some professional fighters name Jesus get knocked out before the end of the fight in US boxing matches. In January 2024, Jesus Trujillo of Miami US, was sentenced to 14 years in prison after pleading guilty to conspiracy health care and wire fraud. There are no (and never will be) Samoan men name *"Iesu"* that I know of and I always say **"in the name of Jesus Christ our savior from Bethlehem and Nazareth"** at the end of all my prayers. Maybe we should say "Yeshua" (Hebrew for "The Lord Saves") so there is no confusion?. You decide.

A problematic issue that appeared in the past three decades is the misguided faith of some Christians which resulted in the death of several children. Some of the parents in some countries refused to take their sick children to the hospital but preferred to just pray for healing. Several parents in the US were later jailed when the children got sicker and several children later died. This is not faith; it is stupid faith. This issue came up during the COVID pandemic in Samoa, where some people relied on prayer and traditional medicine and ended up getting sicker. My dear readers, the scriptures don't offer material prosperity, only spiritual prosperity, and it also doesn't offer physical healing, just spiritual healing. Again, Christianity is wholly a spiritual phenomenon.

Growing up in the 1960's, I would save every *'sene"* (cent) to later buy a ticket to see mainly cowboy movies at the *Tivoli* theater in Apia. I had thought that Indian Chiefs with their colorful headbands were just for the movies and not real people. In early 2000, I attended a workshop on Climate Change in the United States and was one of the presenters. About four Indian Chiefs from various tribes on mainland US participated and were wearing their traditional colorful headgear, as was seen in the movies that starred famous actors like John Wayne, Audie Murphy, Dean Martin and others. My grandchildren don't believe there are such people as Indian Chiefs with colorful headgear, and I understand completely. However, I had talked with these people, shook their hands had some discussions with them and NOAA should also have some reports on this workshop. Atheists don't believe the Gospels but the authors who wrote the Gospels got stories from people who were there in the Holy Land talking and living with Jesus. Samoan cultural history, like those of the Native American Indians, were mainly handed down through our oral story-telling stories and we generally believe these accounts. The westerners also have their own routes of documenting their history, and some have been debunked.

Having read and watched several episodes of twisted and deranged killers, I had wished I could watch justice being done as fast as the legal process allows. Maybe in a few weeks while the horrific scenes are still clear in my mind. Mostly to make me feel at ease and with some satisfaction- it is also a psychologic means of mind cleansing. Sometimes, I think the Last Judgement mentioned in the Bible seem to be too late, and as an anxious human I would yearn to seeing justice being done during my lifetime. That feeling however, is against the divine schedule in the Bible. However, two "instruments" have come to the rescue for impatient people like me. The first device to satisfy this itching arrived a few years ago in the field of Genetic Genealogy which has solved many homicide cases dating back 30 to 50 years. The second, is kind of barbaric (not Christian) device where pedophiles and other murderers are bludgeoned to death, in US prisons, by other inmates, when they found out what these prisoners were in for. One prisoner remarked that repenting and saying I'm sorry is not sufficient. I kind of like this ex-inmate. Please ask your pastors, or better yet do your own research and seek assistance from the Holy Spirit about the theology regarding this concept

Uncaring and arrogant scientists are advocating humans' eventual migration to the moon and subsequently to other planets. First of all, the problem of space junk needs to be addressed before any mass migration to outer space takes place. Recent estimates indicate that there are about a million objects larger than half an inch floating around Earth at speeds of about 17,000 miles per hour. I wouldn't step into a spaceship knowing that there are these dangerous objects floating around in space that will damage any space craft. This phenomenon started in 1957 with the launch of Sputnick 1 and the orbiting debris will increase in the future. The indifference portrayed by these scientists are sickening. These people have not been to the slums, ghettos, and isolated places in several countries like India, Pakistan, Philippines and other poor countries where thousands of people are actually living without electricity, running water and little to no food, not to mention the unforgiving climate and unhygienic conditions. Decision makers, like politicians, should consider these three concepts when examining this concept: Urgency, Empathy and Prioritizing.

These smart people may consider contributing to the many current problems on Earth like cleaning up the Yamuna River in India or providing alternative places to relocate the poor people of the Pacific islands displaced by Climate Change. There is also the decades old problem of nuclear waste and the few unsolved mathematical problems. This concept is similar to one of the four Noble Truths in the Buddhist religion, called Dukkha and the idea of solving Earth's problems, first echoed by professor Thomas Seyfried. They may also assist in worthwhile projects that benefit humans now living on Earth and generations to come, like raising chickens in laboratories. In case they didn't know, there are also hundreds of hungry children in the United States (the greatest country in the world today) not to mention hundreds of homeless people including veterans who fought in wars that subsequently enable them to comfortably continue their science researches. Readers may also be interested in knowing that just recently, US authorities discovered homeless people living in caves along the Tuolumne River in California. According to the US Census Bureau, 46 million Americans live in poverty and this is the third-highest poverty rate amongst developed countries, ahead only of Turkey and Mexico, and 1 in 4 children live in poverty in the US. Using millions of dollars for space exploration while millions of people are in dire situations right here on Earth shows a lack of sympathy and empathy for the indigents, vulnerable and the unfortunate people of this planet. The selfish nature of these scientists portrays the general viewpoint of the privileged including many false televangelists. They live decent and comfortable lives but lack compassion and empathy, but I understand their situation. **They are atheists.** Since they are also world-renowned debaters, they are welcome to debate me on this issue. but they first need to go live with these indigent and unfortunate people for 12 months to digest what I've just said. Good luck.

Space exploration has provided some benefits to mankind but humans can survive without these new technologies. Other fields in the sciences will come up with similar discoveries. When funds are available to communities vulnerable to climate change, would they prefer to build a new museum or opt for new coastal barriers?. Communities needs to address real and pressing problems first. In regards to our Pacific surroundings, we are in dire need of assistance in relation to Climate Change impacts and our neighbors in the Marshall Island, for example, are in dire need to find a way to safely remove nuclear waste from the "Dome".[106], real actual problems right here in our neighborhood.

Sometimes "everyone should be treated the same" concept include kings. During this period, Greece's former king was buried as a private citizen (AP. Becatoros, E, 1-11-23). Ponder on this: Jesus was however, buried in a tomb belonging to a "millionaire".

In 1990, the burial place of High priest Caiaphas was discovered. The remains of Pontius Pilate and several other people mentioned in the Bible have also been discovered. These are only a few of the many relatively new archeological discoveries that proved the existence of people and places mentioned in the Bible. (Professor Paul Maier. Iowa State University presentation).

Unfortunately, several books and presentations by so-called "scholars and researchers" who are atheists, have already been published and distributed to the public claiming these people and places did not exist. I expect public apologies from all these people, but knowing their arrogant behavior, and the certain public humiliation, I don't think any apology is forthcoming.

106

In the US around 2019, numerous conspiracy theorists bombarded the general public with several ridiculous theories, many of which were related to the US presidential elections. If these theories were really true, its followers shouldn't be in prison when they acted on these theories. Unfortunately, many ended up in jail, proving that these conspiracy theories were iniquitous, erroneous and downright misleading. In the US, one leader of these cults[107] was ordered by the court to pay millions of dollars to those affected. Even when these liars are publicly debunked, these "truthers" continue to spread misinformation. Everyone has the right to express their opinions but it is the stupidity of the followers that amazes me. For example, even after reading in the news that pastors have stolen money from their churches, blind followers continue to be part of these churches. For example, a well-known minister in Baton Rouge and New Orleans, for more than 30 years, admitted to defrauding his church (First Emanual Baptist Church) of almost $900,000 but parishioners continue to be members of this church. With the increase availability of resources available on every topic, from now on, I'm not going to blame cult leaders and false Prophets. The followers are to blame. Consider the following:

There is currently a man in Australia name Alan John Miller who claims to be Jesus Christ and I'm not surprised that the number of his followers is increasing. His mother once took him to a psychiatric ward. A similar cult leader name David Koresh in the US led his followers to death in Waco some years back. Additionally, there is another fraud Jesus in Siberia[108] name Vissarion, where he had setup his Church of the Last Testament in 1990. There are definitely lots of lunatics in this world like these fraudsters. It would be very interesting if we could put all these people who claim to be Jesus Christ in one room for a debate on who is the real Jesus. I predict that a fight will break out and they will kill each other...maybe.

107

108

To the Christians, this phenomenon is a serious cuss and this False Christianity claim may have unfairly injected reincarnation into their faith. Absolutely disgusting fraudsters. Many US January 6 insurrectionists, staunch followers of a former President, were jailed during this period. I thought that if Jesus is planning to come soon (before I die) I would volunteer to be His Worldly Coordinator because I know exactly where to get loyal and unwavering people who would go to jail and even die for a belief (or a leader). The places to go are the prisons around the US. where the insurrectionist are jailed.

There is this Samoan saying: *"Ua ui le ma'i"*, meaning there is no cure for the illness. People, many of them intelligent and smart, will continue to agree with and be brainwashed by false preachers, politicians and conspiracy theorists and there is no cure for this phenomenon.

Similar events had happened with some celebrities in the US like Leah Remini, who finally left the Scientology cult after several years. Interested readers can get related information from ":A Piece of Blue Sky" by Jon Atack or from "Barefaced Messiah" by Russell Miller. Thousands of poor people in Oregon in the 1980's continued to follow the Cult of Osho, even though they knew their cult leader has amassed more than 50 Rolls Royce. The cult leaders are not stupid, the followers are.

For Christians, this is why I'm promoting the idea that the public should do their own research of the Gospels and not rely solely on preachers, priests, pastors and other people to interpret for them. The young Samoan generation can also assist their parents to use resources available on the internet. Current new technologies offer multiple avenues and resources to do this, and always remember that the Holy Spirit is **available 24/7**. God's will was channeled through the prophets in the Old Testament and later through Jesus Christ and the apostles in the New Testament. Now it's time for the internet (be very careful of AI) with the help of the Holy Spirit. This process is the method to be used to attain what Jesus tried to explain in his explanation of "light" in John 11.

Mid 2022 underscored the US Roe vs Wade debate related to abortion. The Bible doesn't directly address this topic (Howard A. Melanie. Assoc Professor of Biblical & Theological Studies, Fresno Pacific University. 7/20/22). but scholars have used various texts from the Scripture to argue the Pros and Cons relating to the reversal of the US Supreme Court decision. I personally believe that since abortion is directly related to human life, we need to just follow the will of God who gave us life. Unfortunately, no human being can correctly interpret God's will. Since American Samoa is a US territory, this legal decision by the US Supreme Court may have some implications to American Samoans. In the US, court decisions on critical issues like gun control and gerrymandering, will affect everyday lives of the public, including parishioners of all churches. How about this: "increase awareness and drives for programs seeking private donations for scholarships using household incomes as basis". I have a little understanding regarding this suggestion since I work for an entity that uses household income to determine eligibility to our free legal services.

Bringing back decedents to life is not just a Biblical topic, it has been a scientific issue for many years. Bodies, sperms, eggs and brains of several individuals have been preserved in very cold containers in case some scientific breakthrough comes around in the future that would enable these deceased people to be brought back to life and be healed by future discoveries. Recently, scientists were able to reanimate dead cells in pigs. (Farahany, Nita A. NBC News. 8/3/22).

That is, they were able to restore dead pigs cell functions and heartbeats, of pigs that have just died. Imagine what this new development can do to humans in the future as it **blurs the line between life and death**.

Some overseas wars in the past were fought because of divergent religious beliefs. For example, the war between the Catholics,

Orthodox and Muslims in Bosnia killed many people and went on for an extended period, all in the name of religion. According to one of the chiefs in our village, there is a solution to this mess, and that is for all the people of that country to agree to disagree and respect other people who have different beliefs and religions. This concept would have undoubtedly saved thousands of lives in the Balkans conflicts. He said this is a tough and complex solution but it's the **only one**. Rome wasn't built in one day.

There is this question that has come up for many years and I've yet to find a worthy answer until the writing of this book. The question is: Is it God's fault that murderers kill so many people including friends, relatives, pastors, women, the homeless and children?. After years of research, I've never come across a worthy answer to this question from overseas sources. Now let the Samoans provide a wholesome answer. After the arrival of the LMS missionaries, the village *faifeau* were trained by LMS missionaries on Biblical concepts and ideas to be imparted to the youngsters at village LMS Sunday schools. One of these was the memorization of a few answers to basic questions like: "*o ai ne faia oe?*"- who created you. The children were told to memorize the answer: "*O le Atua*" – God. This was the case and the conventional answer for many years until the last decade when Pentecostal Pastor Iliafi preached that the answer should be: "the parents", not God. He said that a child is created by the natural parents, not God. I totally agree. Take for example, the case of serial killer Jeffery Dahmer. Mr. and Mrs. Dahmer came together and produce a son (a very disturbed one) named Jeffrey. It wasn't the parents' fault though. God created Adam and Eve, not Jeffery and ALL other murderers. So, it is crystal clear that God didn't create serial killers, it was their parents. Cain was given freewill in Genesis and he decided for himself. Therefore, people shouldn't blame God. Of course, **God doesn't need defending**, so ponder this answer, keep studying the Bible, pray, fast or come by to have a conversation with me at my banana plantation.

Samoans should now adopt a Christian view on life as soon as possible since there is no repentance in heaven where Jesus won't be a savior but a judge.

There is a continuous relationship between Israel and Samoa. In May 2023, the Ambassador Extraordinary and Plenipotentiary of the State of Israel presented his credential to the Independent State of Samoa.

The idea of virgin birth has been dismissed by atheists-scientists for decades as they fiercely argue against this unfeasible concept. Well, in January 2018, a female crocodile in a Costa Rican zoo laid a clutch of eggs (New York Times. Greenwood, Veronique. 6/7/2023) which was peculiar since she has been living alone for 16 years. In an incubator an egg from the clutch perfectly formed a stillborn baby crocodile. This was reported in the journal Biology Letters by a group of researchers who wrote that the baby crocodile was a pathenogen – a product of a virgin birth, containing only genetic material from its mother. This phenomenon has been identified in king cobras, sawfish and California condors. This brought up the possibility that "pterosaurs and dinosaurs might also have been capable of such reproductive feats". Samoan reef fishermen should know that some fish found in the reefs, like the parrot fish, can change their sexes, so when the pastor eats his fish during the Sunday to'na'i, he would be eating either Jack or Jill!.

There's an interesting story that depicts human beliefs and their perspectives on life. The story goes: Mr. Refrigerator had asked Mr. Electricity what would happen if someone unplugs the cord. Mr. Electricity replied that he will return to where he came from hundreds of years ago, and according to physics, he will not die. On the other hand, you Mr. Refrigerator, once the power cord is unplugged, you're done. Christians are like electricity while atheists are like refrigerators.

Atheists, when they need help, should ask for assistance from other atheists since they have the same beliefs. They shouldn't ask the theists. A dying man, a communist, asked his son to go and ask his brother (the son's uncle) for some money to purchase some medicine as he was in so much pain. The uncle was a successful businessman and obviously could afford to pay for the expensive medicines but refused to give any money to his nephew. He told his nephew that he is as capitalist and doesn't believe in providing money to communists because they will not repay his money but might give it to the state.

With recent discoveries, interpretations and data analysis relating to Biblical scriptures, there are now controversies as to the location of a new Temple that is supposed to be built before the return of Christ. Some researchers have concluded that a new Temple should be built on the area known as the City of David. Others say it should be on the area of the Temple Mount. As I wrote in my previous book, the new temple maybe interpreted as a concept of preparing peoples' hearts to receive the second coming of Christ. The foretold temple therefore would be "peoples' hearts" not a physical temple structure. Furthermore, the continuous political and religious conflicts in the Middle East would prohibit the building of such structure. One might also argue that there is no need for a second physical temple in relation to Revelation "But I saw no temple in it, for the Lord God Almighty and the Lamb are its temple". Sometimes, theologians should listen to voices from the pacific islands.

There is an interesting social aspect regarding execution. In Oklahoma, it costs about $70k to care for an inmate annually and this would be about the same amount to pay for two school teachers for the whole year. Prioritizing and then making a decision might be a good exercise for our readers. You, the reader can figure out where this analysis is going or maybe just decide to prefer to continue to cater for a serial killer or prioritize for the children's education. You decide.

When millions of humans agree on something based on verifiable evidences, it's hard to persuade or caution them otherwise. An international athlete who had just won a gold medal from the Olympics, had his/her photo flashed all around the world; had the event verified by high tech equipment and several professionals is in **fact** a true and undisputed world champion. However, a few years later, the achievement can be invalidated if steroids are found during drug testing. I can pass a test by cheating, but did I really pass that test?. Does evidences (or lack thereof) and validated data (or peer-reviewed papers) really provide undisputable truths?. Not always. Faith in Christ doesn't rely on any of these man-made platforms and paradigms because these are relatively foolish and infantile, and of course Christianity is mainly spiritual.

A downside to requiring formal peer-reviewed scientific publications to support atheists' arguments can be portrayed by a "supposedly scientific breakthrough" that emerged several years ago when it was announced that someone had discovered the Missing Link mentioned in the Theory of Evolution. It was later discovered that this was a hoax. Unfortunately, a few scientific papers have already been written quoting this discovery. There goes the reliance on scientific reference. Astonishingly, the Missing Link is still missing according to Dr. Frank Sherwin. Furthermore, recent cautions by a few scientists on the use of "voodoo science" on US national television, underscore this serious concern. There is a relatively new phenomenon stating that some people with blue eyes may eventually have brown eyes sometime during their lifetime. (Jakes "You've been lied to about genetics:.SubAnima. 2023). Darwin's Theory of Evolution is **NOT** etched in stone up to now. Readers may also examine Dr. Jeff Tomkins presentation "Debunking Evolution and Proving Creation" and several related presentations on the ICR Discovery Channel.

Requiring authoritative evidences like reports from reliable and confirmed scientific sources does not always mean that the information is true. This approach has been used by atheists for many years to argue against Christians. An article about the famous Tiananmen Square Chinese Tank Man who blocked a line of Chinese Military Type 59 tanks by standing in front of the tanks during a demonstration in June 5, 1989, appeared in the London Evening Standard newspaper, soon after the famous demonstration. The event went viral all around the world. The newspaper attributed the article to a professional journalist name John Passmore. A television interview with John Passmore several months later revealed that he was definitely **not the author of the article**. There goes the conventional reliance on standard "reliable" sources often used by atheists.

As for poof, our Christian God doesn't need to be proven. Proof is a human concept not a Creator-God-supernatural concept and is also a man-made science platform that man created. To prove the creation or existence of any physical object, e.g., planet Earth, evidence need to be from **<u>outside that object</u>**. To prove intelligent creationism, evidence needs to come from outside the universe, and for the Christians, this proof would come from their God who is both inside and outside of planet Earth. No human being can scientifically prove the existence of the Christian God because the diminutive human brain is incapable of such a colossal undertaking. Christians should also stop getting into the trap that they need to prove to the atheists that the Christian God exists, because they scientifically can't- maybe they can, philosophically, but then atheists would tear their explanations apart. The God of the Old Testament exists because HE exists: Alpha (beyond the Big Bang) and Omega (afterlife). HE exists beyond time, space, energy and matter. A concept that physicists are claiming they are very near in solving. The elephants have bigger brains than yours and scientists will never solve these problems.

Other issues that need to be explained to the atheists include:

i. The Christian God can do whatever HE wants because HE is God, and this include allowing non-believers who blaspheme to continue to live.

ii. HE also doesn't have to have a reason nor explanation nor scientific proof for anything.

iii. As for the concept of reality, this is what God created and man has been trying to describe (as oppose to explain) through science for many decades. Science is good and has also helped with Christianity. For example, all the technology that assists with the spreading of the Gospel, like radio, internet and television. Christians must stop trying to explain reality by bringing in God to every situation. God created Earth and man is here as steward (Genesis 1) but he has slowly damaged the environment (refer to the section on Climate Change) and killed one another (World Wars, abortion, murder). Bad situations on Earth are mainly man's doing with a **little**[109] help (just a little) from the devil. **Nature is reality** and science has continuously tried to explain most of its grandeur up till now but will finally come to a stop in the next five hundred years when science stalls and limits of human knowledge is baffled by HIS heavenly craftsmanship. It is comforting to believe in something good but it's really uncomfortable (according to some atheists) to exist in the negative atmosphere of the unbelievers 24/7.

To further explain (a): One of my friends had four vehicles in 2022 but he sometimes walk to and from work and also to the market and nearby stores. The means and resources are there but he just didn't feel like using those. God has the power and resources to do whatever he wants but it seems that He doesn't utilize (or intervene) these all the time. Maybe He wants humans to learn important lessons on their own regarding life, like promoting justice and equality (including LGBTQ rights) across all tribes. Furthermore, ponder the following (World Video Bible School. 2020 AP Apologetic Press.org) from various scholars.

- Matter demands a maker
- Life demands a life-giver

- Design demands a designer
- Intelligence demands and intelligent creator.

There are also various proofs of Bible scriptures from outside the Bible. For example, readers can examine the "Doubters Guide to Historical Evidence Outside the Bible"- Apologetics Guy – Dr. Mikel Del Rosario. Examination of presentations by "World Video Bible School" can also shed some light from another perspective.

Taking the stories in the Bible literally have sometimes result in disputes between these two approaches. But this shouldn't be the case if people understand that faith doesn't require matching of scientific evidence to realities. Televised debates between these two ideas are very interesting and informative but the public should understand that there will always be a disagreement between these two schools of thoughts, until science runs out of explanations. If at the start of a debate between these two approaches, the moderator would ask the parties that they need to present evidence-based arguments. Immediately, the Christians would be at a disadvantaged since they will have little to nothing to present, unless they argue from the perspectives of creationists like Ken Ham or Jason Lisle. An interesting observation is that scientific evidences are sometimes proven wrong in later years and sometimes hoaxes and planting of evidences occur in real life. Atheists relying on these approaches should be careful. Christianity on the other hand doesn't change and the Christian God never changes (Malachi 3:6).

Ernest Renan's Book "Renan's Life of Jesus" which had some information related to these issues were later discredited by the discovery of the Dead Sea Scrolls. His book was not only anti-sematic but also depicted Jesus as an ordinary man. Later, author Albert Schweitzer (and many others) in his book "The Quest of the Historian Jesus" refuted Renan's determinations. One of the greatest scientists, C. Ptolemy (150 AD) had a theory on creation which was later debunked by Galileo. How would one's children feel if in later years their atheist father is seen all around the world as a liar. I have a solution. Just keep quiet for the next 1000 years, and maybe new discoveries in science might help you out.

During a debate between Bill Nye, a renowned scientist, and creationist Ken Ham in 2014, e Bill said **he didn't know the answers** to most of the questions like: How does conscious evolve from matter and where did the "items" that started the Big Bang came from. Ham invited people listening to their debate to visit the Creation Museum near Cincinnati to get more information regarding the arguments he claimed during this debate.

Since scientific methods require that experiments to prove something need to be **repeated or replicated** by other scientist (s); how would this be done for the events that occurred millions of years ago?. The scientists said that they rely on assumptions and projections, but then these are NOT facts, just theories but they demand facts from the Christians. Maybe the scientists need to re-define the scientific processes/theory?. A very serious concern regarding Gregor Mendel's experiments (the cornerstone of genetics) was recently discussed by some in the genetic scientific community. It noted that when Rafael Weldon tried to replicate Gregor Wendel's genetic experiments, **he couldn't**. There goes the fundamental of genetics and points out the positive need to require replicating experiments. Samoan students in High School and College taking biology should be familiar with these experiments and should read on this subject to gain a true perspective of genetics. Further information can be gathered from "The Genetics Pedagogies Project" and the book "Gregor Mendel's 200 Year Anniversary" by Amir Teicher.

Atheists who mainly rely on science for their arguments should wait until all the unclarified notions regarding time, space, energy, matter and gravity are well rationalized and proved, so that their final pieces of life's puzzle would appropriately all fall into place. An example was noted in a recent podcast that claimed that "no one has ever correctly measured the speed of light" and that there was a paper claiming that some researchers were able to measure the speed of light. Later, some scientists provided a paper debunking this claim. I thought that with all the brilliant atheist-scientists available on Earth, this seemingly simple problem would have been tackled years ago! It goes to show that scientists are thousands of years away from answering all questions regarding creation. (The Christians' last piece of life's puzzle fell into place in Jerusalem about 2000 years ago). This suspension may also give their field of work time to explain the problems with the Theory of Evolution and conclude the Theory of Everything, which will then provide sufficient support for their arguments. The String Theory also has problems according to Dr. Brian Keating (2023). The James Web telescope discoveries have also brought up more questions regarding the Big Bang Theory and some scientists have also noted that there are 16 wrong predictions in the Big Bang Theory. Data from this telescope seem to "debunk many modern theories of the universe" according to EYES 200M "James Webb Telescope Just Debunked all Modern Theories of the universe. 2023". This presentation also suggested that the more curious we are about the universe, the more questions we come across, including questioning the Standard Model. Cambridge's Quantum physicists Peter Russel's' new book "Let it Go" may show some light to this issue. People can watch the evolution of Quantum Mechanics and see the changing theories, the debunking of others and the continuous changes that emerge during its progress.

Readers should note that there are also other related theories besides the Big Bang Theory. According to Dr. Jim Tour., the following texts can also provide related information.

1. "The Origin and Nature of Life on Earth". Erick Smith and Harold Novowitch.
2. "Evolution – Our Understanding of the Origin of Life"- James Shapiro
3. "The Emergence of Life". Pier Luigi Luisi.

Recently, there was been a paper (BIGTHINK, Answers in Genesis. "Evolutionists Belief Shattered by New Theory") suggesting that there might have been, not one but two Big Bangs. Guys, please make up your minds before debating. A local deacon now suggests that there might have been three Big Bangs to correspond with the Trinity concept. He also added that "nothing created everything" concept from the perspective of science is ridiculous. I don't believe this man because he never attended college, but he's a good farmer though.

"Scientists have not come close to explaining the origin of life". According to scientist Neil deGrasse Tyson, and humans know only 4% of what's out there". I believe that science will not and can't explain everything. This issue is discerningly explained in Professor John C. Lennox in his book "Can Science explain everything". For those that need a philosophical approach to the issue of the origins of the universe, Dr. Paul Nelson's discussion of paradigms should help. For those still doubting scientific evidences confirming creation, there's is a very powerful discussion by Dr. Jason Lisle in the video "Astronomy Reveals Creation" in which he describes the four secrets of the cosmos that confirms the Bible. Additionally, Gary Parkers "Creating Facts of Life: How Real Science Reveal the Hand of God" may provide further understanding. In science anyone, muslim, Christian, atheist and others, if they do the same experiment and run it the same way, they should get the same results. Atheist and scientist William Baines stated that "Origin of Life field as a whole is unconvincing, generating results in Toy Domain that cannot be scaled to any world scenarios" and that "organic products from origin-related experiments and procedures suggest that the building blocks of life is utterly irrelevant and misleading". According to Professor Jack Szosta the origin of life codes protein still needs to explain the substrate for protein synthesis, the genetic code and the function of the first peptides. Science is still struggling. Guys, take a break from debates and concentrate on your experiments.

Jermy England, a Jewish rabbi in a related discussion stated that: "you start with a random clump of atoms, and if you shine a light on it for long enough, it should not be so surprising that you would get a plant!. Also refer to Psalm 102:25 "God laid the foundation of the earth, and the heavens are the work of your hand".

The capsule carrying specimens from an asteroid touched down in Utah in September 2023. The sample from this project may shine more light on the origins of Earth and other related phenomena. This underscore my opinion for atheist to **wait** until "all their ducks are lined up correctly" then proceed to debate the Christians who have already got their arguments ready from Genesis 1. According to Dr. Stephen Meyer, the book Darwin's Doubt stated that there are problems in this theory examined in peer-review literature but not reported to the public. There is also the question regarding the Cambrian Explosion where Darwin himself wrote "To the question why we do not find rich fossiliferous deposits belonging to these assumed earliest periods prior to the Cambrian system, **I can give no satisfactory answer**". There are some theories on this subject but they're just that, theories.

With the emerging theory of quantum mechanics, one would think that it will bring better explanations to a variety of questions. However, According to British scientist Roger Penrose, quantum mechanics is an inconsistent theory and also reminded us that Einstein also said that quantum mechanics is incomplete. Scientists have been arguing for years that you can't "create something out of nothing". Well in 2022, quantum physics scientists started experiments that **suggest** (it's just a suggestion that hasn't been proven) "a strong electromagnetic field can rip particles and antiparticles out of the vacuum itself, even **without any initial particles or antiparticles at all**". An orator working on his taro plantation and who didn't finish High School commented: How can one rips particles that shouldn't be there in the first place?. It seems that the scientists now have changed their minds and now agree that the universe can be formed from nothing. They might argue that science is still in progress and may come up with contrary information. Well how about utilizing their immense knowledge in solving real world problems instead of pondering the origins of life which they surely will fail. The theory on the origins of life however, was suggested by Julian Schwinger nearly 70 years ago. Unfortunately, laboratory experiments like this suggestion, are not the same as real world situations where the conditions can be markedly different. Interactions of various parameters[110] in nature with the complicated surrounding environment can result in different results from laboratory experiments. According to Dr. James Tour, science is not getting nearer to explaining the origins of our planet. In fact, he said, science is getting further away from explaining it. Calculating when t=0 in the Big Bang Theory is a good start. Readers can also view the documentaries by Professor John C. Lennox, a mathematics professor from Oxford University, like his "The Logic of Christianity", to get more insight to these issues.

As of the date of this publication, scientific explanations regarding the Theory of Evolution **are not sufficient.** Dr. David DeWitt, Professor of Biology wrote: "Mutation and Natural Selection are inadequate to account for the amazing molecular complex integrated systems found in living cells". According to Answers in Genesis Cand. Answers TV. "evolution is simply fake". Furthermore "everything you know about genetics is wrong" according to a presentation by Adam Rutherford -U10 Realfagsloiblioteket. So where does evolution stands?

Other scientists also noted that "natural selection makes sense, except in the light of population genetics", in a relatively new concept of Genetic Drift and Neutral Evolution. Researches of the 2022 Nobel Prize winners in physics recently dismissed Einstein's phenomenon famously called "spooky action at a distance" and I believe that a few more theories maybe dismissed while the poor atheists-scientists struggle to argue the impossible. In 2022, scientists discovered a new blood type labelled Er in a UK hospital. I listened to one US televised debate and blood types were mentioned by atheist-scientists.

110

Not all scientists are entirely against the concept of God and creation. Newton, one of the most brilliant men that ever lived once invoked the **possibility of a creationist**. Several elite scientists also believe in the notion of a Christian God. An interesting statistic from Astrophysicist Neil deGrasse during this period noted that 40% of US scientists were classified as religious scientists but only about 7% of these (classified as elite scientists) believe in God. This figure **should be zero if science is actually against the idea of a God**. (More than 40% of the Germans believed in the Nazi propaganda which killed millions of Jews). Again, atheists should wait until all their ducks are lined up correctly and they also need to decide if they believe in science since some of them disagree with top scientists during debates on US National television – confusion. Evidence can also be time (period)-dependent. Scientists also claimed that hundreds of species have gone extinct but some, for example, lions were deemed extinct in Chad Sena National Park, then 20 years later, researchers got a photo of one beautiful lioness in the area.

The most famous atheist in television debates, Richard Dawkins when asked about a topic related to DNA, seem to contradict himself when he said that he believes in some kind of creation but it would be from outside our universe, from a more advance civilization and not created by the Christian God. Readers can do their own research (it's also on YouTube) on this unreasonable and naïve character. He also stated that Religion is dangerous to mankind. Well, maybe other religions are but definitely not Christianity. If Christianity is removed from Earth there will be so much chaos and death that the condition of this planet will be in contrast with the Garden of Eden in Genesis. His friend Christopher Hitchens **seem** to agree with this notion of chaos if Christianity is removed from society when Christopher mentioned the book "When it was Dark" by Guy Thorne.

Science will continue to struggle with increasing questions regarding the origin of the universe for another 1,000 years. Why am I saying a "thousand years"? This seems too long a period considering the fast progress of science. Well, according to scientist Sabine Hossenfelder, Particle Physics have been making wrong predictions and postulations in the past fifty years regarding Supersymetry, Proton Decay, Dark Matter, WINPS, Axions Unparticles, and no one has seen these predictions in a laboratory.

Christians don't have to wait for any new discoveries. They know where they came from and where they're going. So, my atheists friends, please step aside, (continue debates on other topics besides 'origins") and stop polluting the airwaves with deranged ideas. According to one Gilbert K. Cheserton: "The problem of disbelieving in God is not that a man ends up believing nothing. Alas, it is much worse, He ends up believing anything". For atheist Richard Dawkins who advocated that his wife loves him and he has proof, he should watch the scores of court cases where the husband later discovered, after many years of marriage, that his wife has been having an affair behind his back for many years. Sir, take the advice from your friend Neil de Graase Tyson **"be careful of your senses".** Lastly Dawkins believe that animals were actually thinking of managing their energy between fight/flight response and conserving energy for things like laying eggs. No evidence, nice try. Readers may find the book "Freedom, the end of the human condition" interesting as it relates to this topic as it suggests that human savage and aggressive behavior is not due to genes but to conscious mind-based psychological behavioral condition. (Professor Harry Prosen and Biologist Jeremy Griffith).

Related to this issue is a quote from Richard Conn H: "a theistic view of our existence becomes the only rational alternative to solipsism".

There are many other videos, podcast, presentations and articles available on the subject of Christianity and atheism and the public should seek these out instead of viewing the same old Dawkins, Tyson, Hate preachers and others. The character of Dawkins however, is beneficial to Christianity. Readers can refer to Dr. Alister McGrath's book "Coming to Faith through Dawkins" where the author explained how this staunch atheist accidentally led people to Christ. I'm not going to give Dawkins a bad label even though he once said "It is absolutely safe to say that if you meet somebody who claims not to believe in evolution, that person is ignorant, stupid, insane or wicked, but I'd rather not consider that" (New York Times 1989 quote). I forgive him because he was just referring to his family, relatives and friends.

Some atheists stated that Christians use the God concept to explain things (some call these "gaps") they can't be explained and that scientific explanations are more preferable since it utilizes scientific evidences and facts. This is completely incorrect and that many people like myself welcome scientific explanations most of the time, and I had also explained the weaknesses regarding evidences on another Section of this book. Scientist often pointed out that "in the absence of any evidence we can postulate, where there is a possibility ......", so these are **their "gaps"**.

The Bible is not only the text that has been interpreted by so many scholars. One scientist recently noted that the "survival of the fittest' concept in Darwin's theory may also be interpreted as the "survival of the luckiest", the scientist then proceeded to provide his supporting arguments which were very interesting. Furthermore, there is also a new related concept called "sexual selection" that has come up. The explanation is that humans select their mating partners, not the environment nor through the result of "survival mode". Therefore, the offsprings would not be the result of natural selection. An example is the preference by many men of beautiful women in Russia. Additionally, Federico Faggin's book 'Artificial Intelligence vs Natural Selection provides another related theory. You the reader can now see the enormous amount of additional and contradictory theories that have now inundated the minds of so many people.

This is my personal viewpoint regarding the issue of creation: Science can't explain any phenomena before the Big Bang – this is where the Christian God comes in. Science can't explain the afterlife – this is where the Christian God comes in. In between these two points, science (especially physics) is welcomed to explore and at the same time provide useful discoveries for mankind like in the past where several inventions were utilized, and recently AI (Artificial Intelligence) all exist between these two points. Between these two points lies reason, rationalization, religion and science; areas that the human mind can examine and assess to their heart's satisfaction. Before and after these two points are the areas that the relatively insignificant human brain can't comprehend. An example of such concepts beyond these two points is the question: "who created God?- Section A below. Atheists should therefore refrain from being definitive about concepts beyond these two points, which are beyond human comprehension. That is also why I believe Dawkins's book "The God Delusion" was written by a delusional man. One of the references in this book was John Lennon of the famous band, The Beatles. John Lennon apparently used drugs and became delusional and probably causing conflicts with bass player, Paul McCartney._(an aside: I consider myself one of the best Bass Players in the Pacific). Science is not just about evidence but also how people interpret evidence. May I suggest to readers to also read the book "Return of the God Hypothesis" by Stephen C. Meyers. Atheists should acknowledge that man is both physical and spiritual - humans have souls and I've discussed this topic in another section. Therefore, the human condition should be explained in both realms. This idea maybe similar to the **holistic** approach in health advocated and **proved** by the 102-year-old Dr. Grady McGarey. Her book "The Well-lived life" will explain this issue.

During a televised debate between Shaldir Ally and Douglas Jacob there was a childish comment that Jesus should have clearly stated that He (himself) is the eternal Son of God and because that concept is crucial to Christianity, it should have been explicitly written in the Gospels -maybe in Red Letters as being done in some Bible versions. Here's a lesson for this juvenile perspective. Last year I send my grandson and granddaughter (Gideon and Finiana) to a

Church event and told them to report back to me what happened and what the pastor and the deacons said. Later they described the whole event and what was said. A few days later I asked my daughter about this same event and I found out that my grandkids forgot to mention some of the important messages and announcement made, including a $10 donation required of the members of our *Mafutaga a Tama*, due the following day. So, I missed donating. I'm not going to explain the idea I'm trying to relay in relation to the injudicious comment mentioned above. I know anyone reading this section is an adult and has completed High School, so they can figure out why the "expected" statement from Jesus is **missing** or they may talk to my grandchildren to explain the concept. To further explain this issue, there is a concept called "selection" in examining old transcripts where the authors would "select" certain events and also not monitoring all the related details of events. For example, Mark mentions Jesus healing a blind man while Mathew wrote that two blind men were healed. The two stories should be taken as contradictory but may be seen as a result of 'selection". Readers should also be aware that authors of old scriptures sometimes use paraphrasing and sequencing in their writing. Paraphrasing being the communication of the same meaning in different words and sequencing is the chronological/topical order of events. An example given by some Biblical scholars: Mathew unites the cursing and withering of the fig tree to keep the material on the same topic together. The authors of the old scriptural transcripts didn't have recorders so they opt to just give us the gist of the matter. Therefore, since the gist of the scriptures are correct, the gospels have achieved what they set out to do.

Salani's Cosmos Model of Life

Section A: **BEFORE THE BIG BANG**	Section B: **PLANET EARTH**	Section C: **AFTERLIFE**
God in Genesis; before the Big Bang.	Reason, Science, Human Conscience and conscious man-made Religion, Hypnosis, Reality, Meditation, music, poetry, love, moral, integrity, psychic phenomena.	God in the Book of Revelation, Human Soul, Human Spirit,

- "God in Genesis" is not in Section B, even though He can and had intervene in the past. ALL the problems on Earth (Section B) are man-made, so God shouldn't be blamed for any events like Climate Change, genocide; hunger etc.; humans can understand all the elements in this section, and will continue to improve his understanding as science progresses. This is the simple section.

- Humans can't, and will never be able to comprehend Sections A and C because it's beyond the human brain capability (also refer to my Facebook question to 2 top scientists). Readers interested in hypnosis can enjoy discussions by expert David Spiegel.

- Astrophysicists and other scientists are currently trying to describe (as oppose to explain) the wonders of the universe, but they will eventually fail because of the inadequacy of the small human brain and the infinite unexplained forces, energy, and other phenomena not yet discovered scattered billions of miles away where humans will never be able to reach. It will take millions and billions of years to reach just a few of the known planets, so humans need to have babies after babies born inside a capsule before reaching and examining these planets and also assessing if the laws of physics on Earth are the same there. I don't think so because our fastest jet fighters can't even catch up with UFOs (refer to well-documented events on the internet). Additionally, humans won't be able to lie on a beach (if there are any) without their heavy astronaut gear, or have intimate relations with these heavy gear on!. Lastly, how can we reach many other planets that are racing away from us?, the resources and science available to this planet are not sufficient to accomplish this. Not in a million years.

• Note that the Big Bang Theory is just one of a few theories including the interesting Conformal Rescaling. Additionally, these is also a theory of The Fourth Geosphere (Dr. Eric Smith. Earth-Life Science Institute. Sante Fe Institute) suggesting a more ecology-centered biology. It is fruitful that atheists declare **before** any debate, that since science is progressing, it is possible that they maybe completely wrong when new evidences are discovered. But then, scientists admitting that they were wrong is so rare that I found only one example of a scientist's admission of being wrong during my three years of research for this book. It is therefore not surprising that Dr. Neil Turok, a world-renowned scientist of the Perimeter Institute for Theoretical Physics stated that "most of what theorists do is wrong, and that its very important that scientists publicly admit they were wrong when they find out later their theories were false. To be fair to these scientists, theories are just that, theories.

In my research for this book, I read that a renowned scientist Dr. Terry Mortenson agreed that everything can be explained by time, space, chance and laws of nature; evidently all of these were created by the Christian God. This statement is very similar to my reasoning below. As for the question why God created Earth. Ponder this: God was creating other planets but was not satisfied with these, so he created his masterpiece, Earth and consequently gave His only Son for the redemption of its inhabitants. If readers are not satisfied with this standpoint, how about this: After God created other heavenly bodies, he began to **enjoy** doing it, so he created Earth then rested (on the Sabath) and he saw it was "good" (Genesis1), then decided to continue creating other planets.

These factors could offer some explanations to the general question "why did this happen". In a broad-spectrum, I humbly offer my personal three-phased principle regarding life's events in general: "**Maybe God meant it to be ...there's a logical/scientific explanation ....because He is God**". This is actually an extension to my previous perspective of life in general (the "Toy Train" community) offered in my previous book. Try the first part of this updated text to answer your personal questions relating to local social/world-wide issues, and if you're not satisfied, try the second then the third part. You will be surprised how my philosophical principle can actually shed a light on the various questions about various societal and religious issues regarding life. Even atheists and Adventists can utilize the middle and third part. To clarify part of my viewpoints, visit local places like *Mapuifagalele* and Hope House, or even overseas hospitals like Saint Jude's, Shriners Hospitals and hospitals for the mentally challenged. Observe the residents and feel the truth and warmth in my viewpoints. Additionally, the atmosphere and scenes in these places had made me a believer in euthanasia. Remember that God and Nature are under no obligation to make sense of reality, but there's no harm in trying to rationalize it. Seeing and experiencing life in general, one can conclude that its either a good experience or it's misery, depending on one's location, circumstances and situations. Sometimes, someone else (like a drunken driver or a murderer) is in charge of one's destiny unknowingly; it doesn't matter if one does everything right. The world, in reality, is not perfect nor logical. People in high places and those with money often dictates a community's progress and goals.

A Samoan lady left her young children alone at their home as she needs to go to work at the Star Kist cannery (in American Samoa) since she was the sole breadwinner of the household. Something bad happened to one of the children and the women rushed back to her home. A relative and police asked her where she was and why she left the children alone at their home. On a clear and fine night while watching the millions of stars in the pacific skies, I had the feeling that the Christian God is still **out there working hard**. Did I just answer the question: Where was God when all the misery in the world is going on?. **Maybe HE has to go to work and that HE had already provided man with brains to fetch for himself; all the resources on Earth to survive and that humans have free-will to decide?**. Readers may think that I'm being ridiculous but my view is backed by science. Ponder the Inflationary Theory, which explains how universes are being created all the time. Readers can research this topic to confirm my viewpoint.

In 2023 a Connecticut women had to leave her children at home while she left for work. A fire later broke out and her four children died. Should she have stayed home and starve with her children, or does she has to go to work to feed her children?. According to one local untitled man from Pago Pago village: "Grow up spoiled brats, (I think he meant humans) stop complaining, and if life is not so good, give it back, other animals can use it". I'm not sure I fully agree with my friend Suki. Has any human stayed around 24/7 watching over their children, even joining them inside their school rooms? or even inside the toilets and bedrooms even when they are adults?.

A village chief Taala, left for a funeral in New Zealand after he had built a nice house and bought a car for his children in the village of Salani. After a year, he returned and found the car completely wrecked and the home in disarray. I made a comment to him one Saturday that if he had stayed behind, forget about the funeral, other relatives in New Zealand and the Samoan traditions regarding funerals to look after their home and children this wouldn't have happened. Can you readers catch where I'm going with this? The Creator just staying around looking after the spoiled brats, (sorry I meant humans); and people like me blaming the chief for not staying around?

In recent years, some people in these islands began to rate and assess churches according to the resources they are required to give. These evaluations then become the basis of their decisions to join a certain denomination. These decisions had no spiritual basis and therefore the health of the churches shouldn't be evaluated according to the number of parishioners.

I also believe that there are other reasons and explanations to most things in life. It's just that humans have not found all of these in the eyes of science (the complete rationizations is in the Bible though). As world-renowned astrophysicist Neil deGrasse Tyson noted, "science has continuously tackled the mysteries of the universe one by one". Science has gradually tried to describe (as oppose to explaining) various phenomena which were once believed to be the work of the devil. For example, a person with epilepsy was once believed to be a demon possessed case.

Some of the literature available to students in the US are bias and several scholars have suggested these be revised to provide a more impartial portrayal of facts and reality. For example, when it comes to the topic of inventions, US students rarely read about Dr. Shirley Jackson, Lewis Latimer, Marie Van Brittan Brown, Otis Boykin, Lonnie G. Johnson, Charles Drew an and many other **black** scientists who made significant contributions to scientific inventions. Recently a black entrepreneur name Moses West, showed off his "Atmospheric Water Generator" machine which has produced clean drinking water from the atmosphere, and has used this machine to provide clean water to the people of Puerto Rico after hurricane Maria and also for the military.

This issue was also in mainstream US media during this period when school officials in Florida debated the concept of Critical

Race Theory, which seeks to understand how racism has shaped US laws. The Gospels, written several years after the events mentioned happened, maybe inclined to some bias or imperfect portrayal of actual events. But again, I personally believe that this was all included in HIS divine plans. Readers will find out this is true when they get to heaven. Unfortunately, there's little chance I will be there since people who like *Vailima* are prohibited from entering this kingdom. But this bad habit is not that evil according to three arguments suggested by some of my professional musician friends:

1. Our uncle Noah in the Old Testament behaved in a similar manner.
2. Jesus turned water into wine.
3. The ten commandments did not disallow drinking alcohol

I hope the readers don't take these arguments from these sinners seriously.

Epilogue and In Memoriam

Man has and never will "invent" anything from scratch.
Everything that has ever been built by humans was built using "elements" (sub-atomic particles, energy, forces, waves etc.) that were already on Earth, and created by the Christian God. Humans (like Dawkins and Hitchens) came to Earth with nothing (Job 1:21). They can confirm this statement by asking their birth mothers. Remember, scientists agreed that energy can't be created nor destroyed". This is true because the Christian God already created energy in Genesis 1. When atheist Richard Hawkins was asked about the concept of creating something out of nothing, he started his reply by referring to some particles/forces – **which is something**. An absolute stupid reply.

Dictatorships and communism which have always resulted in the killing of thousands of innocent people and the destruction of valuable structures including churches, will continue to survive until countries are no longer ruled by dictators, who according to history, are all mad men. During this period, the example was the invasion of Ukraine. Some democracies like Samoa, has been ruled by a democratic form of government **but in reality**, was ruled by a 'dictator', according to some Samoa legislative representatives who left the HRPP Party after recent elections. Fortunately, there have been no mass killings or indiscriminate destructions in these islands mainly because of our unique culture and the weighty impacts of Christianity.

The following Poem depicts a Samoan Chief/Deacon's contribution to his village community and his church.

Pondering the Chief

The old Samoan chief with an unfinished sennit, "afa" by his side
Stares towards the sea as the afternoon tide comes in;
The village fishermen were late arriving from their morning troll (toso)
His short white hair, ever blinking eyes under burly white eyebrows and
*   crouching posture*
Revealed old age and nothing more
His shaky right hand still holding onto his "ali"; his pillow for the last
*   fifty years*
He might be a wise old chief ... or lazy elderly, a challenged senile
*   commoner ...?*
pondered the noisy children playing on the "paepae"
But one can only guess, since he hardly speaks
The men of the village watching him from a nearby "fale" contemplated ...
He has led the village cricket team and won many championships for
*   many years*
He had built the biggest fale talimalo (Samoan guest house) in the
*   middle of the village*
He had been the village's main "tautai" for the village's va'a alo fishing
*   fleet for decades*
And also cultivated the biggest taro plantation in the district
Then it slowly dawned on the people of the village ...
The old guy, a cultural deiform just sitting there, staring at the sea ...
Is an accomplished village idol,
measured by the tasks he completed, the duties he performed,
the charges he realized, the conflicts he triumphed
but mainly by his adoration of his newly-found Christin faith

I wondered out loud ... as I look at the deacon's portrait on the wall
and inquired if Finiana Tamasailau has come home from school.
(Fini Aitaoto)

History has documented various noteworthy philosophies or principles by foreign philosophers and I think this is an opportune time to add a local perspective regarding life from this south pacific location, that relates to the subject of this book. I personally believe that Intelligent Creation by the Christian God is fine in regards to the initial emergence of humans, but with the consequent utilization of free will and various factors, life's journey and general human final destinations are determined by a myriad of elements like drunken drivers[111]; a microscopic virus e.g. COVID 19; politics[112]; hard work (Sheng Thao was living in her car and sleeping on strangers' couches with her newborn son, unsure where she'd find her next meal".- during this period she became the youngest mayor of Oakland, California); persistence (American Samoa's congresswomen Amata Coleman finally won the local Congressional elections after more than 10 years of struggle – I was one of her supporters all this time); drugs, mental illness; chance/luck[113] and various other circumstances and situations. Life is a vivid garden with a multitude of colors, mimicking nature.

When the human experience and knowledge completely fails to explain or give reasons for any event or condition, that is the exact point where the concept of the Christian God comes in, **not as a "gap" but a necessity**. Religion is different from spirituality. Christianity, to me, is a purely spiritual phenomenon and also according to Jesus Christ himself (John 18:36). Therefore, it doesn't guarantee a fulfilling and happy physical life without tragedies in this physical world, like having cancer.

My wife, a very avid Assembly of God parishioner who spent

111
112
113

most of her life attending to church matters and hailed by her pastor as a saved soul who has so much faith in Jesus Christ, passed away in 2021 from cancer. Good works, belief in God plus faith in Jesus Christ didn't save her, so the explanation by some pastors that it's these qualities or factors that should save one's life is **absurd**. Maybe they should clarify that they meant spiritual life. Additionally, her faithful pledge of her tithes for more than 40 years should have "kicked in" like Employment Retirement plans (e.g., US Social Security) where retirees receive benefits from saved funds **at their time of need**. Expected assistance, like healing, should be received **in time of need** but it didn't because tithing doesn't work that way. The usual suggestion of "not putting all your eggs in one basket" comes to mind. It was unfortunate that her basket's bottom gave in, after years of being a faithful servant of the AOG church.

This concept can also be illustrated in a story about a Catholic Church employee, Ercole Orlandi, who served under three Popes and was the person in charge of the Popes' audiences. His daughter, Emanuela was kidnapped and there were clues that the Pope and the Vatican knew about the kidnapping. Because Ercole faithfully served under three Popes for many years, he hoped that he would get some information regarding the kidnapping of his daughter from the Pope himself. He was wrong. Just before he passed away in 2004, he told his family that "I've been betrayed by whom I served faithfully". A similar event also happened in the island of Manono in 2023 where the Methodist Church disallowed the funeral service of a young women because she was in a de-facto relationship. According to the Samoa Observer, the father of the young women was very disappointed with this church since her daughter had served this church for many years, serving in the choir and was also baptized in that church. Three comments can be made regarding this incident.

1. The pastor's old and worn-out Bible is missing the pages of John 8.

2. This is also true of the pages of Mathew 7. So, people shouldn't condemn the poor pastor.
3. Samoans don't have to wonder what it will be like during the Judgement Day. They can just go to this church at Manono and get a feel of it!.

Apparently, I refused to bring God into my personal situation but since I'm human, I still feel devastated and reading Psalm 88:18, "Lover and friend hast thou put far from me, and mine acquaintance into darkness". I'm at times confused. I wish I was like Abraham, the "Paradigm of Faith", but I'm not.

Speaking of cancer, my wife decided to stay away from chemotherapy and radiation because of its cruel effects. Metabolic therapy would have been another choice but it was too late, and I think it won't work, besides she said "Ill just leave it to God".

Anyway, Christians should understand that any privileges and benefits from being a devout Christian can't be "withdrawn" on Earth. Christians are rewarded only when Christ returns (Mat 16: 27) or when one gets to heaven. Maybe this is one of the reasons why millions of non-Christian folks hesitate to join the Faith and resulted in others leaving. Realistically, disadvantaged and oppressed folks who are constantly in dire conditions, need assistance **right here, right now** on Earth, not after life! Documented interviews with scores of Ukraine people displaced by the Russian invasion in 2022 would help explain this. Indigent and destitute people in overseas places like the Middle East, the Philippines, India, Africa and other places, are often lured into terrorists groups not because they agree with their doctrines, or because they are immoral humans, but because these terrorists provided food and other necessities **right there and then.** If my family runs out of food and other necessities and there are no other sources to obtain these from, I would join these groups, but refrain from killing. I value the lives of my children, relatives and friends more than politics and any religion, including Christianity. Most pastors would disagree with me because most of them live comfortable lives. Some local churches have increased the number of new recruits after applying the same concept - right here and right now. Practical circumstances and obtaining local truths from experiences can help intending new churches.

Again, Christianity is a spiritual concept with its rewards claimed when Jesus returns, not during one's immediate need. What about the millions who are not Christians?, ask your pastor. Also, many famous Evangelists overseas like Tammy Baker, who died of cancer in 2007, reflect my explanation. More personally, my LMS missionary parents were seriously ill and broke most of their lives, but I believe they had very healthy Christian (spiritual) lives. Sometimes as I grew up, I thought this was not fair, maybe or maybe not. I also believe that the concepts of Predestination and Justification by Faith are purely spiritual and shouldn't be interpreted in a physical manner. Try walking through the poverty-stricken slums in India, Philippines, Pakistan, certain parts of the US, the Reverton dump in Jamaica, observe the human tragedies, extreme poverty, filthy living conditions and one would experience a humbling feeling. Try to apply the above two concepts. They simply don't work and don't make sense in a physical manner. Maybe Ephesians 1:2 – 6/11 can help.

My situation is therefore, a much less heartbreaking situation than those faced by millions of people around the world who have multiple tragedies within their communities and families (or maybe I'm not a selfish person by nature). It's natural to be sad or devastated by such events but at least I'm a little comfortable because I think I understand **just a little** regarding this phenomenon. I now understand grief well and it's so personal that it sometimes deceives people looking at me. The truth and effects are often hidden but one has to move forward in life because we shouldn't dwell on things we can't change. It's hard but it's the only viable way forward. I hope you the reader would read this and hopefully my rationalizations would help you in your time of sorrow. Please let me know (before I croak) if this had helped you. My wife's passing had reminded me of a perspective that "if one finds a true soulmate, and shared a heart, and when one passes away, the other has the obligation to live life in full for both. Dear Di, I'm still here and writing for both of us.

But did I pray earnestly to God to save my wife?. Yes, I did with all my heart and faith, yet she left me. But at least I understand that God did answer my earnest multiple prayers (at least 100 per day perhaps). Sometimes I think because God is God, He dosesn.t have to answer any request. For those who believe God answers prayers, here's my two cents.

I think God answers prayers in the following ways.

1. God may answer NO
2. He may answer YES
3. He may answer WAIT
4. He may answer: Do this first.
   [This concept is similar, but not the same, to the statement by Columbus Bishop Earl Fernandes that Catholics who want to enjoy corned beef and cabbage on St. Patrick's Day should:
   • Make a pilgrimage or visit any church in the Diocese of Columbus named in honor of St. Patrick.
   • Assist at Mass at any church, chapel or oratory on March 17.
   • Pray the "Breastplate of Saint Patrick" a prayer attributed to St. Patrick.
   • Engage in a "pious devotion" such as the rosary, the stations of the cross or eucharistic adoration.
   • Perform an act of comparable penance on some other occasion during the third week of Lent, ending March 18.]
5. He may answer OK (different from Yes)
   Pastor Earl Diggins of Forest Home Baptist Church lost his wife of many years and he asked God about his wish that

he needed to leave this world the same time as his wife. The pastor passed away shortly. (Fight of Faith. Doug Eaton)

6. He may answer "sort of no"
   Apostle Paul asked God to heal him of an affliction and God said that "his grace was sufficient for you". 2 Corinthians 9.

7. Or there maybe deafening silence
   A Samoan lady asked his father who was dying with cancer: Why him?. The father whispered to the daughter, why not?. This answer stayed with the daughter, as a silent mystery for many years.

In my case, the answer was NO. If God had answered YES, then there will be no need for a heaven as Earth will be like heaven where everything will be OK, and since everyone is equal in the eyes of God, all similar prayer requests will have the same answer, YES. I am still heart-broken but at least I have **just a little** understanding of how things work on Earth in relation to God by applying my principles mentioned above, to get through life's problems. Praying through Psalm 25: 4 "Lead me in thy truth, and teach me …". had provided me with just a **tiny glimpse** of His ways. As one of the instructors teaching in one of the local Bible Colleges once pointed out "people shouldn't bring God into every bad situation".

Here are some examples of how God answered the prayers of some pastors. During this period: AOG Pastor Joe Amosa asked God to extend his life after the doctors told him he had a few weeks to live. He got an operation and is still living as of the date of this publication. More than 10 years ago, AOG Rev. S. Mageo asked God for a few more years as he was seriously sick. He was given 13 more years, then passed away in American Samoa. Apostle Viliamu Mafo'e asked God for divine healing. He was not healed and passed away in April 2023 after several months of illness. During this period, a pastor in a Pago Pago church had a stroke, asked God for divine healing, but never fully recovered. He later resigned. More than 30 years ago, an AOG parishioner asked God to spare his very sick mother and pledged to serve God if God heals his mother. The critically ill woman was miraculously healed and Pastor Tarrant became one of AOGs dazzling pastors in American Samoa for many years. I'm not considering the medical aspect of these events as I'm only examining the idea of God answering prayers. Martin Luther, before the Reformation, was believed to pray to a certain Saint when he was caught in a severe storm, promising that he will become a monk if he survives that storm in 1505.

Here is at testimony from a presenter (Dr. Gary Habermas, Shoreline Community Church) in the US, related to this matter: "My wife was sick of stomach cancer so the Elders came to the hospital and prayed for her. Four months later she died". The same theologian stated that he and his wife kept a Prayer Journal and after two years they analyzed the data. It turned out that God had answered 67% of their prayer request. I think in my case it was 80%. The rainbow has many different colors, so are the many answers to situations in life.

From a secular humanism perspective, Epstein, the first non-clergyman selected as Chief Chaplain of Harvard[114] during this period said: "We don't look to a God for answers, we are each other's answers.". When the kingdom of Tonga was devastated by the volcanic eruption and the tsunami in early 2022, the churches in the Pacific area and some in other parts of the world played major roles in organizing assistances for the distressed island community. The people, through God's instructions (**but not God himself**) provided the needed assistance. Now let's face reality. I think most people, especially those who asked God for healing when their loved ones are dying would get a "no" answer......

While millions of people around the world live in misery and poverty, the Samoan Islands will hopefully remain the world's most peaceful location[115] on Earth - minus the growing drug problem.

114

115

This problem is like the hope that one day there will be one religion for all the people on Earth. Unfortunately, this will never happen because government officials including the police are involved in drugs. Churches should provide some assistance to this problem because it will definitely affect the parishioners.

Peace is essential to any religion's security and influence, but it can only be accomplished if Samoans embraces and respect its varied established churches and their diverse doctrines (no more new churches please!), continue to practice the original customs (as oppose to the financially-burdened versions now chafing the community), never use the Electoral College system used in the US voting system (it can allow narcissists to lead a nation and who would "mock evangelicals behind closed doors" according to US Republican Senator Sasse); vote in the Attorney General(s), and also seriously ponder numerous truths from private influencers and activists that contribute to the healthy political conversations via social media.

Caution: only a few Samoan journalists are impartial. For example, all the 100+ articles written by "Dear oh Dear" praised the HRPP Party and very rarely wrote anything positive for the FAST Party. In March 7, 2023 Newsline Samoa wrote: "The FAST-Government economic policies since taking over the leadership role have been conflicting and doing more damage than good so far". The childish and unprofessional sunlinesamoa[116] (maybe an offshoot of the US Fox News) crew even mocked the idea of someone referring to the Samoa Prime Minister as President, as though it was the Prime Minister's fault. In August 2023, there was an Editor's Note in the Samoa Observer noting that:

"The Samoa Observer Management unreservedly apologizes to the Deputy Prime Minister, Tuala Tevaga Iosefo Ponifasio for the incorrect reporting pertaining to the above-mentioned story and reservedly withdraws all imputations wrongly made and express

---

116

regret at the error by our reporter concerned. Disciplinary actions will be taken against the reporter for the dereliction of duty as a Samoa Observer reported".

The world of journalism should ponder the daily/weekly expressions of their personal perspectives because this is absolutely unfair. The public should be left alone to consider and ponder on their own regarding any issue. Let them breathe for a few days and then newspapers can publish editorials once every three months, not daily or weekly. Be fair. Despite a statement[117] by US President Nixon that the media is an enemy of the government, I truly believe the opposite. Fair and professional journalism is not only essential for any democracy but also for Christianity's growth. Dear oh Dear should attend the AOG Faleniu church regularly to get some advice.

Some minor practices and church traditions should also be practical and adapt to real conditions. During this period, a pastor attending a funeral service in Samoa, and was wearing a suit, was profusely sweating in the hot and humid environment; whispered to a friend that he needed to go and change to a more suitable attire because of the overwhelming heat. The pastor got up, walked a few yards, collapsed and later died. A thin *elei* shirt with a matching *ie faitaga* and without a tie, would be preferable. Parishioners of various churches are participants of the above events and therefore the church has a critical role in molding the progress of the Samoan society and advocate the needed adaptations to its practices.

A professional and free press is also critical to the fluid progress of local governments and churches. In May 2022, unfortunately, Samoa was ranked 39[118] by the press freedom watchdog, Reporters Without Borders, and the Samoa Electoral Office banned the media from witnessing the official counting of ballots during a by-election in 2023.

Samoan writers are encouraged to write on crucial subjects like

117
118

Religion (in a local context) and Culture and not rely on foreign writers who sometimes get their stories wrong and offending like Margaret Mead's book "Coming of Age to Samoa", which was based on her research in the islands of Manu'a and was later debunked by other notable overseas researchers like anthropologist Derek Freeman, who in his book "Margaret Mead and Samoa: The Making and Unmaking of an Anthropological Myth" challenged all of Mead's findings. Challenging these types of erroneous information about the Samoan Islands should be done by Samoans themselves and this is what I did during a conference in Hawaii circa 2010, attended by at least four Social Science Professors. After my presentation, (described in my second book) a Professor commented that he was glad to hear directly from a Samoan Chief, regarding the Mead controversy. Internet and newspaper articles also sometimes provide erroneous information regarding the Samoan Islands. For example, according to website: https://www.worldatlas.com "Christian instruction is, however compulsory in the public primary schools throughout the country". This is absolutely incorrect. One of the presentations on the internet presented 100 "facts" regarding various topics. One of the facts presented was: "In the Samoan Islands it's illegal to forget your wife's birthday". A relatively new podcast "Breaking the Silence" incorrectly portrayed the Samoan Culture as a culture that refuses to allow the discussion of abuse. Abuse happens all over the world and it's a "world culture item not a Samoan-only element". Councils of the chiefs in the villages, government agencies, churches and NGOs in these islands have all worked together in the past years to address this issue and there has been several positive headway in this field. Regional organizations and the United Nations have also participated in discussions and workshops dealing with this issue for several years. It is sad to see erroneous materials done without diligent research, like this podcast, appear on the internet for the world to see. This podcast is actually the one abusing the Samoan culture. Another ridiculous video "Essential White History", supposedly using false information from one "Emma", provided inaccurate assertions regarding marriage arrangements in these islands. I understand this type of presentation because it's from superficial, shallow, fraudulent and trivial minds. Again, I challenge any of the people who created these presentations to a debate, preferably under my breadfruit tree or in my banana plantation.

During the Hamas invasion (not a war) of Isarael in 2023, a silly podcast by Mark Hill appeared on the internet. The assertions by that podcast were mostly incorrect and misleading according to a detail analysis by Robert Spencer. This conflict has very complicated disputes but the most devastating issue is the killing of many innocent people, including children and people with special needs. The tunnels in Gaza have always been portrayed as evil places where Hamas bring in weapons for their terrorist agenda. However, in a 2023 60 Minutes Documentary, it was clearly shown that these tunnels also bring in thousands of building supplies, household items, medicines and other essentials for the public.

I hope overseas Publishing companies would allow books relating to Christianity in the Samoan Islands to be published and let the reading public decide its appropriateness, and not censor these texts according to their "personal" beliefs. Unfortunately, this nearly happened to my second book, but its contents were not that dramatic as the *Satanic Verses* by Salman Rushdie. This is why I decided to add a lot of Footnotes to this book so Publishers/readers can see the sources and not wholly blame this author for my usual professional and unbiased comments.

Conclusion

My more than 15 years of research for my three books and more than 60 years' experience as a member of a couple of churches, in addition to many related life experiences, observations and interviews, I've come to the following conclusions:

* There is definitely life after death. Demons (e.g., experienced by many Catholic exorcist priests), paranormal and human spirits of the dead, are real. Stories of traditional immortals like *Nafanua, Telesa and Sauma'iafe* are true (I personally saw one of them).

* We have more than enough churches in the Samoan islands and we don't need additional churches, as this will bring more confusion and economic and social problems to our people.

* The incalculable universe out there with its immeasurable features will never be completely explained by the minute human brain. Humans living in outer space is and will always be a myth as common-sense dictates that this enormous reposition when accomplished, will result in the absence of the natural beauty of nature with its magnificent colors and fascinating bio-chemical processes Christians revere each day.

* People should do their own study of the Bible instead of relying on the multitude of false preachers and cult leaders that will lead them astray and sometimes to death. There is an abundant of resources available on the internet, knowledgeable pastors and Bible College instructors (visit Malua and Piula) and of course the Holy Spirit to assist anyone trying to understand the Bible. Samoans should however, continue to participate in their respective churches.
* Samoan pastors should stay away from politics; just stay in you lanes.
* The first Samoan version of the Bible should be the version to be used in church services – God had a plan for that first undertaking by the *Papa o Misi* in doing this first translation. The new Samoan version can be used in Bible studies, Youth Conferences and Bible Colleges.
* Psychic phenomena is mostly true and real.
* God has not spoken to any prophet for many decades. If God has spoken to any living prophet, there should have been be a warning about the Hamas invasion of Israel in 2023.
* Near Death Experience (NDE) is real.
* Not all human lives are precious. Consider the stories from those whose lives have been destroyed by serial killers.

Future generations should appreciate this trail of continuous events and perspectives in this book, which should facilitate their researches and understanding of the progress and impacts of religion, mainly Christianity, on the local Samoan community.

The ink of my pen has again run dry and local readers need to step up and contribute to this home-grown affair which I personally think had been consecrated by the Almighty Christian God. I'm satisfied with life in that I've contributed **just a little** to the Gospel of Jesus Christ. Offering to write my three books regarding the churches can be better explained by the following verse from the

LMS translated Hymn book which tells of how one can offer oneself to do the Christian God's work.

> *Ave lo'u ola ia aoga*
> *Le Ali'e e i au feau*
> *Ave o'u aso ua totoe*
> *E vi'ia ai o Lau Afio*
> Translated:
> *I offer my life to be useful*
> *For you Lord for your errands*
> *I offer my remaining days*
> *To praise thee Lord*

Furthermore, I've crossed out all the items in my Bucket List except the item "attend to Finiana Tamasailau's wedding". I'm like those centennials who were interviewed by some researchers and who commented that they have "no regrets in life". Soifua.

Fui

Bibliography

Aitaoto, Fini. (2012). *Local and regional perspectives, observations and activities regarding climate change in American Samoa*: First Stewards Climate Change Symposium. Washington DC. DMWR. Western Pacific Regional Fishery Management Council (WPRFMC). Hawaii.

Aitaoto, Fini. (2006). *Notes on certain social science issues relating to fishing in American Samoa*. Social Science Workshop. WPRFMC, Hawaii.

Aitaoto, Fini Fuimaono. (2021). *Progress and Developments of the Churches in the Samoan Islands: Early 21ˢᵗ Century*. LifeRich Publishing.

Aitaoto, Nia, PhD, Shelly L. Campo, PhD, Linda G. Snetselaar, PhD, Kathleen F. Janz, EdD, Professor, Karen B. Farris, PhD, Professor, Edith Parker, Tayna Belyeu-Camacho, Father Ryan P. Jimenez. (2015). *Formative Research to Inform Nutrition interventions in Chuuk and the U.S. Pacific*. DOJ 10.1016/j.jand.2014.11.018

Aitaoto, Fini. (2011). *O le Tala Faasolopito o le Ekalesia Fa'apotopotoga a le Atua i Amerika Samoa – History of the Assemblies of God church in American Samoa*. West Bow Press.

American Samoa Government, DOC. 2016 Statistical Yearbook. (2017). Utulei. American Samoa.

Amadio, Nadine (1993). *PACIFICA, Myth, Magic and Traditional Wisdom From the South Sea Islands*. Photographs: John Tristram.

Angus & Robertson Publication. an imprint of HrperCollins Publishers.

Amosa, Maulolo Leaula T. Uelese. (2002). *Fausaga o Lauga Samoa.* Vega II. National University of Samoa.

Barker, Kenneth. Ed. (1995). *The NIV Study Bible*, Associate Editors Donald Burdick John Stek Walter Wessel Ronald Youngblood. Zondervan Publishing House, USA.

Brown, Malaeoletalu L (1997). *H.C. Tuuiteleleapaga Napoleone. Memoirs of a musician and a Composer".* Research Project. American Samoa Humanities Council.

Bureau of Democracy, Human Rights, and Labor. (Sept. 2000). 2000 Annual Report on International Religious Freedom. U.S. Department of State.

Britsch, R. Lanier. *Unto the Islands of the Sea: A History of the Latter-day Saints in the Pacific.* Salt Lake City: Deseret Book, 1986. 348-428.

Craig P., Ponwith, B., F. Aitaoto, Hamm, D. (1993). *The Commercial, Subsistence and Recreational Fisheries in American Samoa.* Marine Fisheries Review. NOAA, NMFS. Vol. 55. No.2.

C. Steubel, Kramer & Brother Herman (1995). *Tala o le Vavau. The Myths, legends and customs of old Samoa.* 1995. Pasefika Press. Auckland New Zealand. Illust by Iosua Toafa.

Discrete News. (1998). 1999-2000 Church Almanac. Salt Lake City:. 267-68, 408-9, 443-44.

Emissions Gap Report 2019. (2019). UN Environment Programme.

Gibbs, Philip. Nd. Papua New Guinea. *Globalization and the re-shaping of Christianity in the pacific islands.*

https://www.samoanews.com/

https://www.talamua.com/

https://www.talanei.com/

http://www.breakpoint.org/bpcommentries/entry/13/30123

Kramer, Dr. Augustine (1994). *The Samoa Islands: An outline of a Monograph with Particular Consideration of German Samoa.* Translated by Dr. Theodore Verhaaren. Vol II. Polynesian Press.

Malua Theological College (2013). *I'u Leo O le Maluapapa: Tala mai le Kolisi Faa-Faifeau i Malua* / News from Malua Theological College; Lomiga XXIX.

Moyle, Richard. (1988) *Traditional Samoan Music.* Richard Moyle. Auckland University Press, in association with The Institute for Polynesian Studies.

National Symposium on Mental Health. Summary of recommendations. Samoa Ministry of Health. 2003. Ed. Percival, Galumalemana Steven., Petaia, Leota Dr. Lisi.

Seventh Day Adventist website: http://sabbathissues.org/

Severance, C. R. Franco, M. Hamnett, C. Anderson, F. Aitaoto. (2013). *Effort triggers, fish flow, and customary exchange in American Samoa and the Northern Marianas: Critical human dimensions of Western Pacific fisheries.* Pacific Science. University of Hawaii Press. Hawaii.

Tanielu, Lonise Sera Tanielu. (2004). *O le a'oa'oina o le gagana, faitautusi ma le tuitusi I le aoga a le faifeau: Ekalesia Fa'apotopotga Kerisiano Samoa (EFKS).* Literacy Education, Language, Reading and Writing in the Pastor's School: Congregational Christian Church of Samoa (CCCS). University of Auckland. New Zealand.

The church of Jesus Christ of the Latter Day Saints. Newsroom. (2012). *Facts and Statistics.*

The Bible Society of the South Pacific (1969). *The Holy Bible in Samoan, Revised Edition,* 2013-3.708M. Korea.

Tomlinson, Matt. (2020). *"God is Samoan"- Dialogues between Culture and Theology in the Pacific".* Pacific Islands Monograph Series 29. University of Hawaii Press.

The Bible Society of the South Pacific. *The Holy Bible In Samoan* 63, Reprinted from the Edition of 1887, 2015-5M, Nabua, Fiji. Korea.

The Holy Bible. King James Version. First Gospel Publications.

www.malua.edu.ws\

www.adventist.org.ws/education

Endnotes

www.ingramcontent.com/pod-product-compliance
Lightning Source LLC
Chambersburg PA
CBHW051439170526
45166CB00001B/43